SELECTIONS FROM
Newton's *Principia*

SELECTIONS FROM
Newton's *Principia*

A Science Classics Module
for Humanities Studies

Edited and annotated by Dana Densmore
Translation and diagrams by William H. Donahue

Green Cat Books
an imprint of Green Lion Press

Manufactured in the United States of America.

Green Cat Books
an imprint of Green Lion Press

www.greenlion.com

Cover Illustration: The Octagon Room, Royal Observatory, Greenwich; many of the astronomical observations on which Newton relied were made here. Engraving by Francis Place. Reproduced by permission of the Pepys Library, Magdalene College, Cambridge.

Cataloging-in-Publication Data:

Dana Densmore
Newton's Principia: a science readings module for humanities studies

   Excerpts of text of Isaac Newton
   Notes and commentary by Dana Densmore
   Translation and diagrams by William H. Donahue

Includes translation, introductions, notes, commentary, glossary, and bibliography.

1. Newton, Isaac, 1642-1727, Principia. 2. History of Mathematics. 3. History of Science. 4. Humanities. 5. Science and Technology Studies.

I. Densmore, Dana 1945–. II. Donahue, William H., 1943–. III. Title.

softcover binding
ISBN 1-888009-26-8
ISBN-13 978-1-888009-26-2

Library of Congress Catalog Number 2004107473

Contents

The Green Lion's Preface

About this Module

Here is Newton's *Principia* for those of us who want to know something of Newton's greatest work, but who may lack the inclination or perhaps the time to work through all the intricacies of Newton's geometrical calculus.

Newton's new conception of how the laws of the universe are structured challenged centuries of received opinion, and laid a new foundation for our "common sense" understanding of our physical world. His claims about the nature of space and time challenge us to think on the deepest level about these familiar yet strange entities. His General Scholium, concluding the work, discusses the relationship between natural science and theology, and makes fascinating claims about what it means to do science. All of these topics are discussed in the present selections without the use of mathematics, and can be read and appreciated by readers with no special background in science.

For those who wish to pursue Newton's science a little farther, this book also contains some of the fundamental steps in the line of reasoning that led to universal gravitation, the crowning achievement of *Principia.* Newton's first proposition in Book I, in particular, uses elementary Euclidean geometry to construct a dazzlingly powerful demonstration about all possible planetary orbits. The line of inquiry proceeding from this first proposition uses proportions in geometry and very simple algebra to lead us quickly to the famous "moon test" (Book III Proposition 4). This proposition, with which this selection culminates, shows that the same force that makes things fall on earth also holds the moon in orbit.

The selections are introduced and accompanied by Dana Densmore's highly regarded commentary, which has been specially adapted for this module from her more comprehensive book, *Newton's* Principia: *The Central Argument* (published by Green Lion Press). Densmore's notes support the direct engagement of the reader with Newton; they assist the reader in coming to his or her own understanding of Newton's world-shaping theories as he actually expressed them.

About the Series

SCIENCE CLASSICS FOR HUMANITIES STUDIES is a series of study modules designed to bring fundamental works of science and mathematics within the grasp of students and other readers without the need for specialized preparation. The series reflects the Green Lion's conviction that an understanding of science, and especially of the classical works of science, is essential for all students of the humanities. Science, no less than poetry or philosophy, is human thought, a response both to the outer world of our senses and the inner experience of our consciousness. The more profound a scientific work is, the more directly it addresses itself to our humanity; therefore, there is much in the greatest works of science that can be grasped without special preparation. Yet too many educational programs find themselves limited by the supposed divide between the humanities and the sciences—the so-called "two cultures."

Further, teachers and institutions who wish to heal this unnecessary fracture have had to confront two discouraging barriers. On the one hand, classic texts of real science are often found to be forbiddingly technical in content and burdened with terminologies either antiquated or arcane. On the other hand, popularizations of these classics insulate students from the actual workings of thought and imagination that classic texts embody. Green Lion Press has addressed this dilemma with the series SCIENCE CLASSICS FOR HUMANITIES STUDIES, issued in slim, inexpensive student editions under the Green Cat Books imprint.

Each volume in the series is a compact, inexpensive presentation of classic scientific and mathematical texts, offering generous but judicious guidance for the reader. We have drawn on our many years of reading these books with non-specialist students to choose selections of real substance, and to provide helps that make the texts accessible while at the same time allowing the original texts to speak for themselves, in their own voices.

Besides humanities students, this series will be of interest to those interested in science but lacking time or expertise to read these works unabridged and without assistance. It will also serve readers who already enjoy a technical background but who may wish to experience more directly the sources of contemporary scientific concepts.

Classic works of science and mathematics, no less than other works of literature, drama, and philosophy, lead us to questions (and answers) that may enlighten or delight us, or may lead us to a new understanding of the multiplex and often conflicting views of reality presented in great scientific works. It is the Green Lion's aim to enable readers not only to observe but to participate in such significant achievements of thought. Other volumes in the SCIENCE CLASSICS FOR HUMANITIES STUDIES series focus on Faraday's *Experimental Researches in Electricity*, Kepler's *Astronomia Nova*, the world systems of Ptolemy and Copernicus, and Euclid's theory of magnitude and number.

Preliminaries:

About Newton, the *Principia,* Scope of this Module, Newton's Mathematics, and Glossary

About Isaac Newton

Isaac Newton was born on Christmas Day 1642 in a remote farming village in Lincolnshire, England. His father had died two months previously and the baby was weak and sickly. Little Isaac was not expected to survive, but he did so and lived to the age of eighty-four, dying in London in 1727.

Isaac's father had been illiterate; his mother, Hannah, was only semi-literate, but the family had a small farm at Woolsthorpe. When Isaac was three years old Hannah remarried and left to live with her new husband, leaving Isaac to be raised by his grandmother on the family farm. Hannah returned to the farm with three new children when her second husband also died in 1653.

Isaac had been attending the village school. From an early age he had an interest in, and aptitude for, invention and mechanics. His many varieties of sundials showed his growing interest in the celestial mechanics they measured, and were sought after by Woolsthorpe neighbors. When his mother returned she had some money inherited from her second husband, and when Isaac was twelve she sent him to a grammar school in the market town of Grantham, where he boarded with the town apothecary. During this time he was known for constructing astonishingly detailed and accurate models of mills and machinery, whose operations he was quick to grasp and analyze.

At seventeen he returned home. His mother expected him to manage the farm, but his mind was always elsewhere, and he proved an utter failure at that task. With some efforts on the parts of both his old schoolmaster at Grantham, who had noted his intellectual talents, and his mother's clergyman brother, who also saw Isaac's possibilities, she finally agreed that he might return to the school to prepare himself for entrance into Cambridge University. In 1661 he entered Trinity College, Cambridge.

In 1665 the plague, which was sweeping across England, reached Cambridge and the University shut down. Isaac returned to the farm and spent two years thinking about mathematics, optics, and natural philosophy, continuing lines of thought he had started at Cambridge. During these years he laid the foundations of much of the work of the rest of his life.

After he returned to Cambridge, his academic career flourished, and he was named Lucasian Professor of Mathematics when he was twenty six. He made discoveries and published papers in optics, and built telescopes. Starting in 1679 his interest in celestial mechanics was renewed, and he began investigations that led to his development of universal gravitation. The story of how he wrote *Principia* is recounted in the next section.

In 1699 Newton was appointed Master of the Mint, to which he brought his prodigious abilities and energies. In 1703 he was elected President of the Royal Society and exerted strong leadership there. Publication of his very important optical and mathematical work followed. Queen Anne knighted him in 1703.

Newton was regarded in his lifetime as a light of brilliance in the world and he was an influential public figure. Through it all, however, he remained rather secretive and even gloomy, and he never married.

The breadth of his intellect, and his enormous powers of concentration, served the world in ways that people not only of his own time, but still of ours, contemplate with wonder and awe.

The Grand Sweep of *Principia* and its Central Argument. Scope of this Module.

In January 1684 a conversation took place between Robert Hooke, Edmund Halley, and Christopher Wren at a meeting of the Royal Society in London regarding the possibility of deriving Kepler's laws of planetary motion from physical forces.*

Hooke claimed to be able to demonstrate all the laws of celestial motion by assuming a power varying inversely as the square of the distance between the celestial bodies.

This "inverse square" relation had been observed in connection with the spreading of light and the action of magnets, and many people, including Wren and Halley, were speculating that it might apply to celestial actions. Wren had discussed the possibility of an inverse square force law with Isaac Newton as early as 1677.

Wren responded to Hooke's claim in that conversation by offering a prize for anyone who could produce a demonstration that an inverse square force law would lead to the motions of the planets described by Kepler. No such

---

* The Royal Society was the most influential forum for scientific ideas in England at that time. It registered discoveries, discussed and debated them, and often published important works. For a fuller account of the writing of *Principia,* see Richard S. Westfall, *Never at Rest: A Biography of Isaac Newton,* Cambridge University Press, 1984), to which work this summary is indebted. Of the several recent biographies of Newton, this is the one that gives the most emphasis to the scholarly side of Newton's life. Background on the writing of *Principia* can also be found in the introductory volume to the Latin edition used in this translation: I.B. Cohen, *Introduction to Newton's 'Principia',* Harvard University Press, 1971.

demonstration was forthcoming from Hooke, at least nothing that satisfied Wren, and the question stood.

In August of that year, Halley visited Newton at his home in Cambridge and mentioned the challenge, asking Newton whether he knew what sort of orbit an inverse square force law would produce. Newton answered that it was an ellipse, and that he had demonstrated it. Halley, excited, urged him to send the demonstration to him as soon as he could.

The first version Newton sent was a nine-page document entitled *De motu corporum in gyrum* (On the Motion of Bodies in Orbit), submitted to the Royal Society in November 1684. It not only demonstrated the planetary ellipses but also showed how all of Kepler's laws may be seen as consequences of physical forces.

It was obvious to Halley that this was a momentous contribution to placing the mathematics of planetary motion on a sound physical foundation. It went beyond the work of Kepler in two primary ways. First, it was universal, not depending on different plans for actions between planets and the sun and actions between planetary matter and the planet itself. Newton was able to show that terrestrial heaviness and the forces that move the planets were a single phenomenon. Second, Newton's system required fewer contrivances and *ad hoc* assumptions than Kepler's. It didn't require reference to imaginary or hypothetical entities. For example, Kepler supposed two powers in the sun, neither of which could be observed other than by their effect on planetary motion. A kind of magnetism had to be supposed for the sun that followed laws different from those obeyed by the magnetism found on earth. The actions Newton relied on could be easily found and tested on earth.

Halley was eager to have the document published. But Newton wanted to develop it further, and urged Halley to hold off on publication until he could re-work and expand it. All his prodigious intellectual energies were consumed with this expansion. The manuscript of *Philosophiae naturalis principia mathematica* (Mathematical Principles of Natural Philosophy),* was initially delivered to the Royal Society for publication in April 1686.

However, Newton still continued to work on the book, expanding it into three books. The final version of Book III was delivered to Halley for publication a year later, in April 1687. Book III applied mathematical demonstrations of the two earlier books of *Principia* to our world and derived from these foundations the principles of universal gravitation and the elliptical orbits of planets (along with many other things, including the motions of the tides and the paths of comets).

Principia was written in a stupendous burst of creative energy. Newton lived during this period like one possessed. He often forgot to eat. "When he

---

* This is the full title of the work most often referred to simply as *Principia* (Latin for "Principles")—the work to which this Module is an introduction.

has sometimes taken a Turn or two [in the garden], has made a sudden stand, turn'd himself about, run up ye Stairs, ... fall to write on his Desk standing, without giving himself the Leasure to draw a Chair to sit down in."*

In such circumstances, the compass of the *Principia* could not be restrained. Among the many dazzling insights for which he sketched out proofs were: the conceptual equivalent of conservation of kinetic and potential energy; general expressions for orbits under any arbitrary force law; expressions for attractions between finite spheres under any force law; attempts to treat the motion of multiple mutually attracting bodies using approximations and qualitative arguments; the motion of pendulums; the motion of waves in water; the motion of fluids (including a refutation of the Cartesian theory of vortices); explanations of tides and of the nutation of the earth's axis and the motion of the moon's nodes; and laws governing the orbits of comets.

Buried within this heap of brilliant propositions is a central jewel, the establishment of universal gravitation and its use to demonstrate the elliptical orbits of the planets, which constitutes the main argument of *Principia*. This module gives us some high points of this central jewel of an argument.

This development not only a supremely important step from the point of view of the history of science, but in addition it is Newton's practical demonstration of his theory of how science could be done in a way that yields certainty, being (as he saw it) purely deductive.

This attempt to give science a logically sound deductive basis constituted a radical departure from Francis Bacon's inductive method, which was very influential at the time. Bacon advocated collecting many and varied instances of the phenomena under study and trying to see patterns among them.

By contrast Newton used minimal experimental data. His main experimental foundations, the "Phenomena" of Book III, were (as we shall see) very far from being pure observations, but they were based on observations and theory generally accepted. Everything was deduced, using mathematical demonstrations, from these few observation-based conclusions about how our world works.

In his "Preface to the Reader" (pages 3–4 below), Newton describes this revolutionary method thus:

> And on that account we present these [writings] of ours as the mathematical principles of philosophy. For the whole difficulty of philosophy appears to turn upon this: that from the phenomena of motion we investigate the forces of nature, and then from these forces we demonstrate the rest of the phenomena. ... In

---

* Humphrey Newton, Newton's amanuensis for five years during the writing of *Principia*, quoted by Richard S. Westfall in *Never At Rest*, p. 406.

the third book, ..., we present an example of this procedure, in the unfolding of the system of the world. For there, from the celestial phenomena, using the propositions demonstrated mathematically in the preceding books, we derive the forces of gravity by which bodies tend to the sun and the individual planets. Then from the forces, using propositions that are also mathematical, we deduce the motions of the planets, of comets, of the moon, and of the sea. In just the same way it would be possible to derive the rest of the phenomena of nature from mechanical principles by the same manner of argument.

This is the Newtonian version of confirmation theory, which is ingenious and is significantly different from both contemporary and modern theories of scientific explanation.

Kepler had previously shown that the planetary orbits are elliptical. Without using Kepler's laws, building only on his own foundations, Newton successfully used his method to derive the elliptical orbits of planets. This is a test of the method, showing that it arrives independently at true results. Newton's central argument provides this test of the method.

Newton explicitly delineates the core sequence, consisting of the following parts of *Principia*. The mathematical foundations for the development of universal gravitation and celestial mechanics are found in the Definitions, Laws of Motion, and some basic hypothetical mathematical propositions, the first seventeen propositions of Book I.

Then Newton suggests that we go to Book III, where he introduces what he calls Phenomena, a small number of conclusions about our world.

Then we are to follow the primary propositions of the application to our universe, especially in the first thirteen propositions of Book III. Here lies the thrilling derivation of universal gravitation: the discovery that the moon is falling just like a rock (or a pendulum bob), that inertial and gravitational mass are quantitatively the same, that every particle attracts every other particle inversely as the square of their distance. Having established these things, he uses them to prove that the planets will move in ellipses, and Book III goes on from there to investigate the moon's motion, comets, tides, and much more. (In the process Newton invites us when necessary to dip back into Books I and II for a few extra propositions and to augment our observation-based data with results from experiments on pendulums.) This is the procedure recommended by Newton in his preface to Book III.*

This module lays the foundations for Newton's core sequence. It includes his definitions and laws of motion, his famous foundational scholia, his amazingly powerful and remarkably simple investigation of the *central force,*

---

* I have followed Newton's recommended procedure in my book *Newton's Principia: The Central Argument,* available from Green Lion Press. This module contains selections from that book.

the kind of force which, it turns out, governs the motion of all the planets. Then it turns to Book III, where we read his Rules of Philosophizing and his statements of the Phenomena, the evidence on which on which all the deductions of the structure of the universe are based. And we end with an exhilarating finale: the thrilling demonstration that the moon is continually falling towards the earth just as a rock falls. This is the keystone of his unfolding of universal gravitation, which extends terrestrial heaviness to the celestial motions.

About Newton's Use of Mathematics

In undertaking the adventure of following Newton's thinking, we want to avoid the distortion of his thought that results from projecting modern notions. But the modern notions are what we know. How do we avoid interpreting Newton through modern eyes if those are the eyes we have? The following remarks may help raise our awareness and help us understand why Newton's approach is not just accidentally different from our modern one.

Geometry

One question that has puzzled readers ever since *Principia* was published is why Newton chose to express his proofs in an idiosyncratic modification of the classic geometry of the ancient Greeks. Newton had already developed his version of the calculus in algebraic form, but chose not to use his "method of fluxions," as he called it, for the *Principia*. Why not?

I'm not going to try to answer that question here; that's for you to explore as part of the adventure of reading *Principia*. But I do want to say a few things about why it is important to read Newton in the mathematical language in which he wrote, rather than translating him into modern notation.

First, Newton, like Kepler before him, had very clearly formulated ideas about the relationship between God, geometry, and the created universe. He expresses these ideas in the "Preface to the Reader" (page 3 below). It is at least arguable that he believed that geometrical proofs exactly match the inner workings of nature as God actually set it up. To throw this out and substitute some alternative means of expression is to reject this astonishing implicit claim out of hand, depriving oneself of the opportunity to consider Newton's take on the ever-perplexing question of the relationship between mathematics and physical reality.

But there were also more down-to-earth reasons why Newton may have preferred geometry. One of these was the discontinuity in the number system: there are infinitely many spatial magnitudes that cannot be expressed in whole numbers or fractions. This fact, which was known to the ancients (it is the subject of Book Ten of Euclid's *Elements*), made it impossible to put the algebraic form of calculus on a rigorous mathematical and logical foundation. It took over a century for number theory to develop to the point where it could encompass the irrationals, and meanwhile strong arguments were raised against the use of "infinitesimals," as Newton's fluxions and Leibniz's differentials were called. The development of a geometrical version of the calculus provided the continuity that the algebraic version lacked. In its geometric form, Newton's notion of limit as expressed in his Lemma 1 and the Scholium following Lemma 11 possesses impressive rigor and sophistication.

Further, there is evidence that Newton believed that algebraic expression of physical relationships obscures nature by lacking referential clarity, in contrast with geometry, which always clearly expresses the physical objects. Equations are mechanical and opaque, while geometry gives us a clear visual and intuitive sense of what we are proving. So in presenting *Principia* in geometrical form, Newton is inviting us, not merely to follow an abstract argument, but to experience directly the hitherto invisible forces that make the universe go.

A warning about ratios and proportions

The practical aspects of Newton's use of Euclidean ratios and proportions are described below. Here I think I need to warn students about the dangers of substituting equations for proportions, and especially of reading our shorthand notation "$A \propto B$" as meaning "$A = k \times B$", where k is a constant of proportionality.

A general problem with using equations is that they assume the definition of standard units for all variables, and this assumption presupposes the whole structure of mathematical physics that Newton was beginning to develop. In other words, it's anachronistic. We must always bear in mind, as we study this book, that there were no such things as "grams" or "newtons" that could be plugged into equations. The ideas of force, mass, and so on, were defined in terms of ratios, and their relationships were explored and developed in the course of the book. To treat them as magnitudes given in themselves is to obscure the way these ideas grew out of Newton's work. For further discussion of the way Newton thought of these things, see the notes on Definition 1, below.

However, there are more immediate problems with substituting an equation for a proportion: it can lead to absurdities that are not present in Newton's proof, but are created by the substitution. For example, in one of his propositions (not included in this module), Newton states that "QR...is as...SP^2." What he means here is that, when comparing two different places P and p on two orbits,

$QR : qr :: SP^2 : Sp^2.$

The two magnitudes on the left are vanishingly small, while the two on the right remain finite. Nevertheless, each ratio, taken by itself, is well-defined and finite, and the two ratios are the same. No problem. But if we try to turn this into an equation,

$QR = k \times SP^2$,

it obviously becomes nonsense, since the constant k must somehow make something infinitely small (QR) equal to something finite (SP^2). For this reason, you must always keep in mind that a proportion cannot be safely transformed into an equation with a constant multiplier.

That is, $A \propto B$ does not mean the same as $A = k \times B$.

Ratio

A ratio is a relationship between two numbers or between two magnitudes. Euclid defines ratio of magnitudes in Book V Definition 3 of the *Elements*:

> A *ratio* is a sort of relation in respect of size between two magnitudes of the same kind.

Note that the magnitudes must be of *the same kind*: areas may be in ratio to areas, lines to lines, velocities to velocities, forces to forces. But Euclid doesn't compare unlike magnitudes, and Newton follows him pretty consistently in this, and doesn't try (for example) to put force in ratio to area. Since ratio is a relation in respect of size, how would we know whether some area were larger, smaller, or equal to some force? How large must an area be before it is twice as large as a certain force? They cannot be directly compared; they have no relation in respect of size.

When we use algebraic equations, on the other hand, we treat what were ratios as fractions, that is, as quotients of numbers. Then we are no longer thinking of the numerator and denominator as magnitudes with a *kind*. Newton sometimes makes use of equations in this way; however, he avoids doing so, because when we do that we no longer have a geometric, and thus visual and intuitive, picture of what we are doing. The manipulation of numbers using algebra is convenient, but it's a bag of tricks that loses for us the reality of the things behind it, at least for the duration of the transformations.

With ratios we speak of the *antecedent* and *consequent* instead of numerator and denominator. In the ratio $A:B$, A is the antecedent and B is the consequent.

Same ratio and proportion

Let's say we have our given ratio. Now suppose another relationship of like magnitudes with respect to size. This may be the *same ratio* with our given ratio. For example, each consequent may be twice as large as its respective antecedent.

We may call this same relationship "same ratio" even if the like magnitudes of the second ratio are of a different kind from the like magnitudes of the first. For example, in the first ratio, one area may be twice as large as another area; in the second ratio, one force may be twice as large as the another force.

Two ratios that are the same constitute a *proportion* and it is abbreviated as "::". For example, if $a:b$ is the same ratio as $c:d$, then $a:b :: c:d$. This is read as "a is to b as c is to d."

Euclid worked out a number of legitimate manipulations that can be done on the four magnitudes in a proportion, yielding new same ratios.

Antecedent and consequent of each of the same ratios can be *inverted*. If the ratios of antecedents to consequents are the same, the ratios of consequents to antecedents should be the same. Thus if one proportion is discovered or assumed to be true, the "inverted" proportion will also be true.

They can be *alternated*, such that antecedent is taken in relation to antecedent and consequent to consequent. One must be careful with this: if the second antecedent and consequent are a different kind of magnitude from the first, alternation will produce illegitimate "ratios" between unlike magnitudes. Newton avoids doing this; when he must alternate unlike ratios he switches to algebraic equations. If the ratios are not ratios of magnitudes but of numbers, then we don't have this problem. Since all numbers are homogeneous with one another, a proportion of numbers remains a proportion of numbers even after alternation.

We can also *compose* a ratio. There we add antecedent to consequent in both ratios and set them in ratio to the original consequents. This yields new same ratios. Early commentators called this operation *componendo*.

Compounding ratios; duplicate and triplicate ratios

A proportion can also express sameness between one ratio and a ratio that is obtained by multiplying two or more other ratios. In Euclidean terms multiplying ratios together is called *compounding ratios*. Euclid doesn't put compound ratio in terms of multiplication, but in terms of relationships between geometrical figures. In Book VI Proposition 23, Euclid gives his geometrical foundation for the operation, letting the magnitudes that are in the ratios be represented by the lengths of sides of equiangular parallelograms. The ratio of the areas of the parallelograms represents the ratio of the compounded ratios.

Newton doesn't try to represent compounding with geometrical figures, despite his overall approach in *Principia* of using geometrical representations rather than ones of algebra or analytic calculus. As a rule he treats the operation as simple multiplication, although of course a multiplication of ratios. Taking the shorthand of algebraic equations would have meant abandoning the connection to reality and grounded intuition.

When ratio $C:D$ is *compounded with* ratio $E:F$ the result is the ratio $(C \times E) : (D \times F)$. Thus to express that ratio $A:B$ is same ratio as ratio $C:D$ *compounded with* ratio $E:F$, we may write

$A:B :: (C:D)$ *comp.* $(E:F)$, or

$A:B :: (C \times E) : (D \times F)$.

Sometimes a ratio is proportional to a second ratio compounded with itself. This is called *duplicate ratio*. So $A:B :: C:D$ compounded with $C:D$ again; or we say that $A:B$ is the *duplicate ratio* of $C:D$.

Since compounding is algebraically like multiplying, duplicate ratio will lead to what we call squares in algebra. In this example, to say that $A:B$ is the duplicate of $C:D$ is, in this algebraic view, the same as writing $A:B :: C^2 : D^2$.

Triplicate ratio works the same way as duplicate ratio, except that, algebraically, we multiply the ratio times itself twice. Both antecedent and consequent become "cubed" in algebraic terms.

Although Newton was able to use algebra with facility, he chose not to embrace it in *Principia,* and it's instructive to note where he chooses to be Euclidean, where he chooses to be algebraic, and what idiosyncratic middle ground he chooses as his characteristic mode of presentation. (A translation that obscures this—as other present-day translations do in converting duplicate ratios to algebraic squares—does a disservice to the thoughtful reader.) Newton's characteristic way of handling duplicate and triplicate ratio is described in the following paragraphs.

Euclid proved that squares (like any similar figures) are to each other in the duplicate ratio of their sides. Newton, following Apollonius, often makes use of this by substituting the ratio of squares on given lines for the duplicate ratio of the lines. However, it is worth noting that Newton almost always writes "square C" or "sq. C," which can mean either the square constructed on line C or C multiplied by itself. He avoids the superscript notation C^2, although he was familiar with it.

Further, cubes, and other similar solid figures, are in the triplicate ratio of their sides. Therefore, a triplicate ratio can be represented by the ratio of cubes on the two lines. And since a triplicate ratio is a ratio compounded with itself twice, it is equivalent to "cubing" in algebra. Newton writes "cube C" or "cub. C", which can mean either the cube constructed on line C or C multiplied by itself twice. Again, he generally avoids using the superscript "3" to indicate the cubing of a quantity or magnitude.

Shorthand notation for proportions: the $\propto$ symbol

Newton frequently works with ratios between corresponding parts in different figures, or different instances of a construction in the same figure as a point moves to its limiting position.

For example, he might have a proportion consisting of two ratios between corresponding parts like this:

$AB : ab :: AD : ad$

This can be read as "*AB* varies as *AD*" and written this way:

$AB \propto AD$

This "$\propto$" notation is always shorthand for a full proportion. Whenever we see it we know that we are actually talking about at least four terms, the first and second magnitudes being in the same ratio as the corresponding third and fourth magnitudes. The ratio of the third and fourth may actually be a complex formula of compounded ratios or multiplied or divided magnitudes (as, indeed, might the first and second). So if we had

$$\frac{AB^2}{Ab^2} = \frac{AG}{Ag} \times \frac{DB}{db},$$

we could say

$AB^2 \propto AG \times DB.$

Glossary

A fortiori means "the case is even stronger for saying that..."

Alternating ratios. See the Preliminaries page xvii.

Anomaly. In Ptolemaic astronomy, the discrepancy between a planet's observed position (with respect to longitude) and the position it would occupy if it were moving with uniform circular motion about a known center. But in the seventeenth century, and for Newton, anomaly was the the angular distance a celestial body has progressed along its orbit measured from the point of greatest distance from the center of forces. For a planet, the angle is measured from aphelion, with sun as center; for the moon, the angle is measured from apogee, with earth as center.

Aphelion. In the path of a body orbiting the sun, the aphelion is the point furthest from the sun. The aphelia are "at rest" if the body returns to the same point at each aphelion.

Apogee. In the path of a body orbiting the earth, apogee is the point furthest from the earth.

Apsides. The line of apsides is the line between the nearest approach of the orbiting body to the center of forces and the most distant point. (For a body orbiting the sun, the line of apsides is the line between perihelion, the point on the path closest to the sun, and aphelion, the point on the path farthest from the sun. The line of apsides of the moon is the line between perigee, nearest approach to the earth, and apogee, greatest distance from the earth in the moon's orbit.) In an elliptical path, with the center of forces at the focus, this line will be the major axis.

Componendo. An particular operation on same ratios resulting in another same ratio. See Preliminaries page xvii.

Compounded ratio. Two ratios multiplied. See the Preliminaries p. xvii.

Conic section. Conic sections are curves that are found by cutting a cone with a plane. A plane parallel to the axis results in a hyperbola; a plane parallel to a side of the cone results in a parabola; a plane cutting through both sides results in an ellipse (if this plane is parallel to the base of the cone it will be a circle); degenerate conic sections such as point and straight line are also possible with other placements of the cutting plane.

Conjunction. In conjunction, the moon is on the same side of the earth as the sun, and is seen as "new." An inner planet in conjunction will be seen as new if it is between the earth and the sun; it will be seen as full if it is beyond the sun. An outer planet in conjunction will be on the same side of the earth as the sun but will be beyond the sun and will be seen full phase.

Conjugate diameters in an ellipse or hyperbola. Take any diameter. Take the tangent where this diameter meets the curve. Then draw a line through the center of the conic section parallel to this tangent. This line will be the *conjugate* diameter (See Apollonius *Conics* Book I Definitions 5 and 6, and I.47.)

One set of conjugate diameters of the ellipse is the major and minor axes. These are the only ones that are perpendicular unless it's a circle, where all sets of conjugate diameters are mutually perpendicular. (See Apollonius *Conics* Book I Definitions 7 and 8.)

Diameter of Ellipse or Hyperbola. A line through the center of the ellipse or hyperbola meeting the curve in both directions. (See Apollonius *Conics* Book I Definition 4.)

Diameter of Parabola. Any line that intersects the curve exactly once. All such lines turn out to be parallel. (See Apollonius *Conics* Book I Definition 4.)

Duplicate ratio. A ratio multiplied by itself. See Preliminaries page xviii.

Elongation. The angular distance between a planet and the sun; or between any celestial body and another about which it revolves. Although elongations are observed and measured from earth, we may (by a suitable calculation) translate any such measurement into the angle that would be observed from the sun, in which case it is called "heliocentric elongation." In Phenomenon 1, for example, Newton reports the heliocentric elongations of Jupiter's moons with respect to Jupiter.

Enunciation. The statement at the beginning of a mathematical proof that articulates what is to be proven or found by the proof. It is followed by the steps of the proof itself, or in the case of the proofs Newton gives in *Principia*, a sketch of the proof that could be made, indicating the strategy without including all the steps.

Evanescent. The size of a magnitude as it, and another magnitude changing with it, which are continuously reduced approaching a limit of zero, vanish. So, for example, the size of an arc traveled by a body may become continuously smaller as the time for its travel is reduced; it is evanescent just as it is imagined to be vanishing—not before, not after, but *as* it vanishes. Alternatively, the value of a ratio of two magnitudes as the magnitudes themselves vanish.

Gibbous. Of the phase of the moon or a planet, greater than half phase and less than full.

In infinitum. Of a process, continued without limit.

Inversion of ratios. See the Preliminaries page xvii.

Natural philosophy. A term that up until Newton's time had been used for the Aristotelian study of motion and change. After Newton used the term for this book, it was never the same again, and was soon synonymous with what we call physics or astronomy.

Opposition. In opposition, the moon or planet is on the opposite side of the earth from the sun, and is seen as full.

Ordinate. Drawn from a point on the curve to a particular diameter parallel to the tangent where that diameter meets the curve. (See Apollonius Definition 5.)

Parallax. Parallax is the apparent angular displacement of the observed body that results from change in the position of the observer.

Perigee. In the path of a body around the earth, perigee is the point closest to the earth.

Perihelion. In the path of a body around the sun, the perihelion is the point closest to the sun.

Planets. Traditionally the "seven planets" were Mercury, Venus, Mars, Jupiter, Saturn, the sun, and the earth's moon. When Newton says "planets," he may include the sun or earth's moon; judge by the context. Newton also at times refers to the moons of Jupiter and Saturn as "planets." Sometimes he makes the distinction by calling the moons "secondary planets" as opposed to "primary planets." Note that traditionally, and in Newton, the earth is not called a planet, although there are places where Newton states general conclusions about "planets" which we have proved apply to the earth.

Principia. Latin for *Principles.* In this Module and generally, short for *Philosophiae naturalis principia mathematica* (Mathematical Principles of Natural Philosophy), the full name of the work to which this book is an introduction.

Q.E.D. is an abbreviation for *quod erat demonstrandum,* "that which was to have been demonstrated."

Q.E.F. is an abbreviation for *quod erat faciendum,* "that which was to have been done."

Q.E.I. is an abbreviation for *quod erat inveniendum,* "that which was to have been found."

Q.E.O. is an abbreviation for *quod erat ostendendum,* "that which was to have been shown."

Quadrature. The angle between two bodies measured from a chosen third point is a right angle. For example, a planet or the moon is positioned such that the angle between the planet and the sun, observed from the earth, is a right angle.

Reductio ad absurdum. A type of proof in which all possibilities other than the one that we want to establish are shown to involve a contradiction. That is, all other cases are reduced to some absurdity.

Refraction. Refraction is the apparent displacement of a celestial body owing to the bending of light by the earth's atmosphere. For example, as light from the moon enters the earth's atmosphere, a denser medium, it is bent downward towards the vertical.

Sagittae (singular, *sagitta*) of arcs lie on lines that originate from a point on the concave side of the curve, and, passing through the points of bisection of the chords to those arcs, extend to reach the curve. The actual sagitta is the segment of this line lying between the chord and the arc. (The word *sagitta* is Latin for "arrow.")

Sesquiplicate ratio. See the Preliminaries for discussion of the powers of ratios. The triplicate ratio is a ratio of cubes; the subduplicate ratio is a ratio of square roots. The sesquiplicate ratio is the subduplicate of the triplicate; that is, a ratio of two quantities each taken to the 3/2 power.

Syzygies. "Syzygies" means "places of being yoked together," namely, conjunction and opposition, when the sun, earth, and moon are in line.

Versed Sine. A sagitta (q.v.) at right angles to the chord subtending an arc.

Vortex, Vortices. Literally, "whirlpool." This word was used by Descartes and his followers for the swirls of invisible matter that they believed filled all space and carried the planets around the sun, the moon around the earth, and perhaps other planets around other stars.

PHILOSOPHIÆ

NATURALIS

PRINCIPIA

MATHEMATICA.

Autore *JS. NEWTON*, *Trin. Coll. Cantab. Soc.* Matheseos
Professore *Lucasiano*, & Societatis Regalis Sodali.

IMPRIMATUR·

S. P E P Y S, *Reg. Soc.* P R Æ S E S.
Julii 5. 1686.

L O N D I N I,

Jussu *Societatis Regiæ* ac Typis *Josephi Streater*. Prostat apud
plures Bibliopolas. *Anno* MDCLXXXVII.

Newton's Preface to the Reader

The ancients, as Pappus wrote, made mechanics of the highest value in the investigation of natural matters, and more recent writers, having dismissed substantial forms and occult qualities, have made an approach to referring the phenomena of nature back to mathematical laws. It has accordingly seemed fitting in this treatise to develop mathematics insofar as it looks to philosophy. Now the ancients established two branches of mechanics: rational, which proceeds accurately by demonstrations; and practical. To practical mechanics all the manual arts look, and from here its name "*mechanica*" is borrowed. But since artisans are accustomed to work with little accuracy, it happens that mechanics as a whole is so distinguished from geometry, that whatever is accurate is referred to geometry, and whatever is less accurate, to mechanics. The errors, however, belong to the artisan, not the art. One who works less accurately is a more imperfect mechanic, and if any could work with perfect accuracy, this would be the most perfect mechanic of all. For the drawing of both straight lines and circles, upon which geometry is founded, belongs to mechanics. Geometry does not teach how to draw these lines, but requires [*postulat*] that they be drawn. For it requires that the beginner learn to draw them accurately before crossing the threshold of geometry, and then teaches how problems are solved by these operations. To draw straight lines and circles are problems, but not geometrical problems. The solution of these is required of mechanics, and once the solutions are found, their use is taught in geometry. And it is the glory of geometry that so much is accomplished with so few principles that are obtained elsewhere. Thus geometry is founded upon mechanical procedure, and is nothing else but that part of universal mechanics that accurately sets forth and demonstrates the art of measuring. Further, since the manual arts are chiefly concerned with making bodies move, it happens that geometry is commonly related to magnitude, and mechanics to motion. In this sense, rational mechanics will be the science of the motions that result from any forces whatever, and of the forces that are required for any motions whatever, accurately set forth and demonstrated. This part of mechanics was developed into five powers by the ancients, looking to the manual arts, since they considered gravity (which is not a manual power) not otherwise than in the weights that were to be moved by those powers. We, however, are interested, not in the arts, but in philosophy, and write of powers that are not manual but natural, treating mainly those matters pertaining to gravity, levity, elastic force, the resistance of fluids, and forces of this kind, whether attractive or impulsive. And on that account we present these [writings] of ours as the mathematical principles of philosophy. For the whole difficulty of philosophy appears to turn upon this: that from the phenomena of motion we may investigate the forces of nature, and then from these forces we may demonstrate the rest of the phenomena. And to this end are aimed the general propositions to which we have given careful study in the first and second books. In the third book, on the other hand, we present an example of this procedure, in the unfolding of the system of the world. For there, from the celestial phenomena, using the propositions demonstrated mathematically in the preceding books, we derive the forces of gravity by which bodies tend to the sun and the individual planets. Then from the forces, using propositions that are also mathematical, we deduce the motions of the planets, of comets, of the moon, and of the sea. In just the same way it would be possible to derive the rest of the phenomena of nature from mechanical principles by the same manner of argument. For I am led by many reasons to strongly suspect that all of them can depend upon certain forces by which the particles of bodies, by causes not yet known, either are impelled towards each other mutually and cohere in regular shapes, or flee from one another and recede. These forces being unknown, philosophers have hitherto probed nature in vain. It is my hope, however, that the principles set forth here will shed some light either upon this manner of philosophizing, or upon some truer one.

[The rest is omitted, since it deals with details of publication that are not of present interest.]

Definitions

Quantity of Matter Defined

Definition 1

The quantity of matter is the measure of the same arising from its density and magnitude conjointly.

Air of double density, in a space that is also doubled, is quadrupled; in a tripled [space], sextupled. The same is to be understood of snow and powdered substances condensed by compression or liquefaction. And the same account is given of all bodies which are condensed in various ways through various causes. In this I do not take account of the medium (if any) freely pervading the interstices of the parts. Further, in what follows, by the names "body" or "mass" I everywhere mean this quantity. It is apprehended through an individual body's weight. For it is found by experiments with pendulums carried out with the greatest accuracy to be proportional to weight, as will be shown hereafter.

Notes on Definition 1

- **Measure and Proportionality: a Question of Relationships**

"The quantity of matter is the measure of the same arising from its density and magnitude conjointly."

For a present-day student it may not be clear here from Newton's wording that this definition is giving us a proportion, not an equation. When he speaks of the measure "arising," this is one indicator of it being a proportion. The stronger and more explicit indicator is the word "conjointly." Another way to express this would have been that the quantity of matter varies as the density and magnitude conjointly; that is, that two quantities of matter are in the same ratio as the ratio of the densities compounded with the ratio of the volumes.

That Newton is thinking in terms of a proportion is confirmed by the first words in the commentary: "Air of double density, in a space that is also doubled, is quadrupled; in a tripled [space], sextupled."

Newton is not telling us here that we will get a measure of mass (in grams or some such unit) if we multiply a body's density times its bulk. Rather, he is telling us how to compare two or more bodies. The ratio of the masses is the same as the ratio of respective densities compounded with the ratio of their respective bulks or volumes. (Compounding is a way of multiplying ratios.)

What does Newton mean by "measure"? How does he measure things? In *Principia*, he measures things by looking at relationships, that is, he compares two things of the same sort and finds that they are in the same relationship as the ratio (or compounded ratios) of some other pairs of things, always comparing things of the same sort.

Unlike what we may be accustomed to in physics textbooks, in *Principia* we very seldom get statements that any one thing is *equal to* some particular other thing or combination of other things, for example that a particular mass can be calculated by multiplying a particular density times a particular volume. This language of equations, so familiar to us and so convenient, is almost completely absent from *Principia*.

This is in part because in Newton's time the use of algebra was just starting to establish itself against the traditional use of proportions. But it was also a choice for Newton. He could have used algebra; indeed, elsewhere (and even in a few places in *Principia*) he did use algebra and the analytic calculus he had himself developed. But he evidently felt that what he wanted to represent was better conveyed by the language of proportions.

Though cumbersome, the insistence on sticking with proportions did two things that are lost in converting to algebra and equations (where that conversion would have been possible).

First, it stayed clear about the meanings of the quantities being worked with. For example, we understand that one distance might be twice another distance, but whatever could be the meaning of dividing a distance by a force?

Second, it kept things in terms of relationships. One might want to speculate about why that might have been a value to Newton, and to watch as one works through *Principia* for other ways in that Newton seems to be seeing relationships as the way to understand foundational things. Keep this in mind for pondering the meaning of the Third Law of Motion and the understanding of the working of gravity itself.

● **Nature of a Definition:** This section calls itself "Definitions." How are we to understand what is said about things described in a definition, as opposed, say, to what is said in something called a law?

This may not be obvious. For example, a student once asked, "Why does Newton define quantity of matter in terms of density when it seems that we know how to measure quantity of matter directly, but we don't know how to measure density directly?"

This question raises several interesting issues, two of which may be helpful to explore here.

First, the question reveals a confusion between a definition and a procedure for measuring the defined item in a laboratory. Definitions in a mathematical exposition describe how terms will be used and do not establish that the thing defined exists. And if the thing defined does exist, the definition doesn't provide the construction or method of finding it. That may need to come in a later proposition, or perhaps in a postulate. How one would go about assigning a number to the quantity of matter in a given thing in the world is not the concern in the definition (or generally in *Principia*). Rather, he is laying out here the relationship between quantity of matter, density, and size.

Another sort of difficulty with this question is that it claims that we know how to measure quantity of matter directly. Do we? We have balance scales that compare weights and spring scales that measure the force we call weight. But weight and quantity of matter, force and stuff, are different things, and we don't yet know the relationship between them. That's one of the important things worked out in *Principia* (see the next note).

This first definition that opens *Principia* invites us to start thinking about just what mass (or quantity of matter) could be on its own. **Question for Discussion.** Imagine that we have no Newtonian physics, and we're trying to sort out these things we're going to be working with. What might quantity of matter be? How would we want to define it, how would we give people a picture of what we will mean when we use the words? Then consider Newton's answers to those questions in this definition.

We have a tendency, living in a post-Newtonian world, to think of mass as something absolute, the concept created along with the matter itself, and that anyone with any other concept or definition must be ignorant or confused. But many ways were open to Newton in conceptualizing what we now call mass. The way he did decide to think about it, the way he defined it, turned out to be very useful, so useful that it's now hard to imagine any other way of thinking about it.

The way to think about these definitions is not to ask whether we would have picked different words (or equations) to describe what we as his beneficiaries understand these things to be. That might be appropriate if he had been given the understanding modern physics uses and was just writing a textbook to codify it. Then we (or some other modern textbook we have read) might have different or better ways of expressing this understanding.

But our opportunity here is to do something much more thrilling and engaging than fine-tuning a textbook. Rather, we can step back and stand with Newton at the moment of formulating what a useful way of thinking about these things would be.

• Quantity of Matter and Weight

"[Mass] is apprehended through an individual body's weight. For it is found by experiments with pendulums carried out with the greatest accuracy to be proportional to weight, as will be shown hereafter."

As Newton will be doing frequently in his commentaries to the definitions and the laws of motion, he is here looking ahead to what will emerge from this work as a whole, and in particular to what will prove to be true when we pull in information about how things work in our world. In this case, as we embark on this investigation, we don't know the relationship between mass and weight, and these experiments he is referring to are part of the development that *Principia* will lead us through. That mass, or quantity of matter, can be "apprehended through weight" must be shown, and it is indeed finally shown, but not until Book III, which makes the various applications to our actual world. The experiments Newton mentions here are cited, and the relationship of mass and weight established, in Book III Proposition 6, which is not included in this selection.

● **Mass and Quantity of Matter:** The word here translated as "mass" is Latin *"massa."* *Massa* is a large irregular lump of something, a bundle or heap. This is in contrast to our modern use of the word mass, which is more technical and abstract. The technical and abstract term for Newton, corresponding to our modern use of the word mass, was "quantity of matter."

The translation in this selection always distinguishes *massa,* mass, from *quantitas materiae,* quantity of matter.

You won't get into any trouble in reading *Principia* if you mentally substitute "mass" for "quantity of matter." But you might want to notice when Newton uses each term, keeping in mind that, when he used the word mass as opposed to the term quantity of matter, he was likely picturing something in its individuality as a lump of matter.

Quantity of Motion Defined

Definition 2

The quantity of motion is the measure of the same arising from the velocity and the quantity of matter conjointly.

The motion of the whole is the sum of the motions in the individual parts, and therefore in a body twice as big, with an equal velocity, it is doubled, and with a doubled velocity it is quadrupled.

Note on Definition 2

This definition is expressed as a proportion, not as an equation. See the first note to Definition 1.

Again, we are invited to think about what quantity of motion might be. **Question for Discussion.** What is motion itself? How might one measure it (not necessarily in the laboratory, but in thought)? One might have chosen to think of motion as the movement itself, maybe measured by the distance traveled in a particular time, or even just the total distance traveled. (In casual speech, we do sometimes use the word in both these ways.)

Newton chooses in *Principia* to use a definition of motion as the speed and the quantity of matter conjointly—that is, the ratio of their corresponding speeds compounded with the ratio of their corresponding quantities of matter. **Question for Discussion.** What does this tell us about the way he is thinking of motion? How is his picture different from other definitions someone might have made?

Kinds of Force (Definitions 3–8)

Newton defines two types of force for us in this section. The first is *vis insita,* or inherent force, given in Definition 3. The second is *vis impressa,* impressed force, given in Definition 4. In Definition 5 he goes on to describe a particular type of impressed force, "centripetal force," the investigation of which is central to this book. He then further distinguishes three different measures of this sort of impressed force, the absolute, the accelerative, and the motive quantities of centripetal force, in Definitions 6–8 and the following commentary.

Inherent Force (Inertia, *vis insita*) Defined

Definition 3

The inherent force of matter is the power of resisting, by which each and every body, to the extent that it can, perseveres in its state either of resting or of moving uniformly in a straight line.

This is always proportional to its body, and does not differ in any way from the inertia of mass, except in the mode of conception. Through the inertia of matter it comes to be that every body is with difficulty disturbed from its state either of resting or of moving. Whence the inherent force can also be called by the extremely significant name, "force of Inertia." A body exercises this force only in the alteration of its status by another force being impressed upon it, and this exercise falls under the diverse considerations of resistance and impetus: resistance, to the extent that a body resists an impressed force in order to preserve its state, and impetus, to the extent that the same body, in giving way with difficulty to the force of a resisting obstacle, endeavors to change the state of that obstacle. Common opinion attributes resistance to things at rest and impetus to things in motion, but motion and rest, as they are commonly conceived, are distinguished from each other only with respect [to each other], nor are those things really at rest which are commonly seen as if at rest.

Notes on Definition 3

• This is the definition of inherent force, saying, in the way of definitions, how the term will be used. The existence of inherent force is postulated in Law 1 where it is asserted that every body continues in its state of motion or rest unless driven to alter that state by an impressed force. Definition 3 and Law 1 are best read in conjunction by anyone attempting to understand this phenomenon.

• The word inertia was not a familiar physical term when Newton used it in this book. It had first been used in a physical sense by Kepler, for whom it meant a body's tendency to come to rest. Newton borrowed the word, but changed its meaning to fit his physics. The basic meaning is "laziness" or "sluggishness."

• We must be especially careful not to project modern textbook definitions or concepts of inertia and force onto Newton's definitions. We're in another world as we read *Principia,* a world we have entered by time travel. Only careful alert exploration will reveal how much of what we find here is what we're accustomed to at home and how much will be different in interesting and even instructive (but possibly subtle) ways. Concepts evolve over time. Words change their meaning. Formulas of modern physics that claim to be "Newtonian" may in fact use different definitions of the same terms. There is no one eternal meaning of such terms as force. Rather, these are concepts that evolve over time. As we read Newton's words, we're at one snapshot of time—a very important moment, but one more than three centuries in the past.

Question for Discussion. What is our own understanding of force? I don't mean to ask what formula we learned in high school physics. What is this thing in the world, in our experience, that we call force? We can ask how we use the word when we're talking about our own personal experience of the world, when describing our everyday experience with material objects and when using the world analogously for psychological, social, or political phenomena. Do we have a direct experience of this thing (or maybe a set of things having something in common) that leads us to a concept prior to any definitions or formulas we might learn in school?

• Returning now to Newton, and looking at the wording of this definition, we might wonder what sort of force this could be. He says that it is that something that is proportional only to body (note that he says in Definition 1 that when he uses the terms body and mass he means quantity of matter). If inherent force is proportional "only" to quantity of matter it is not, presumably, proportional to anything else (and he doesn't mention anything else). But how is it that something proportional only to body comes to act as a force, and not only act in some force-like way, but be *defined* as force?

One notes that not only does he define this using the word force (*vis*), but he goes out of his way to emphasize that it is important to him to so classify it, calling it an "extremely significant" name:

"Through the inertia of matter it comes to be that every body is with difficulty disturbed from its state either of resting or of moving. Whence the inherent force can also be called by the extremely significant name, 'force of Inertia.'"

Question for Discussion. It is worth speculating before we move on about the way Newton is understanding this force of inertia or inherent force, considering in what sense he is seeing it as a force (*vis*), with the possible insights such thinking might yield about what force means for him. A look ahead at the impressed force of Definition 4 might give something to go on. Is there a way in which inherent force can be seen acting in a similar way to the way he says impressed force acts? How is it different? He calls *vis insita* a "power of resisting." How would that be measured?

• If it seems that the more carefully you read and question the text the more puzzles you find, don't feel discouraged or yield to a temptation to fall back on modern formulas or understandings. You don't expect to have figured out "whodunit" in the first pages of a mystery novel. Rather, let that experience of puzzlement awaken a sense of wonder and give you something to ponder as you go along through *Principia*. At this early point we really don't have a great deal to go on. We haven't seen how he will be speaking about force in propositions or even in the Laws of Motion. We will need to stay alert for how he applies these terms in what follows and continue to feel out his vision and intent.

• "...motion and rest, as they are commonly conceived, are distinguished from each other only with respect [to each other], nor are those things really at rest which are commonly seen as if at rest."

Notice Newton's assertion here of the relativity of motion and rest, "as they are commonly conceived." That is, from our vantage point, we can't tell whether things are absolutely moving or at rest: we can only say whether they are moving or at rest with respect ourselves or with respect to one other. He will take up this question of relativity of motion and whether there is such a thing as knowable absolute motion in the scholium following these definitions.

Impressed Force (*vis impressa*) Defined

Definition 4

Impressed force is an action exerted upon a body for changing its state either of resting or of moving uniformly in a straight line.

This force consists in the action alone, and does not remain in the body after the action. For the body continues in each new state through the force of inertia alone. Moreover, impressed force has various origins, such as from impact, from pressure, from centripetal force.

Notes on Definition 4

• This is the definition of impressed force, saying, in the way of definitions, how the term will be used. The Second Law of Motion describes the operation of impressed force; you may wish to read Definition 4 and Law 2 in conjunction, or at least refer back to this definition when you study the Second Law.

• Newton asserts in this definition that impressed force does not remain in the body after the action of impressing.

He is here contradicting the impetus theorists who believed that the force applied to a body, an impetus, remained in the body. This impetus accounted for the body's continuing in motion after the force was no longer being applied.

His new view replaces traditional impetus theories that were current at least from the sixth century. Earlier versions of the theories tended to see the impetus that was added in launching a projectile as gradually expended in the motion. As the impetus was expended, the projectile slowed down; once it was used up the projectile came to rest again.

In the fourteenth century Jean Buridan asserted that impetus from impressed force remained in the body, only diminishing from external resistance or internal tendency to motion in a contrary direction. Aside from being impressed from an external force, it could be augmented in falling, each increment of acquired speed adding impetus. This view of the impetus as physically present in the body was influential through Newton's time.

Closer to Newton's time, the view of force as something that resides in a body and can be transferred to other bodies by impact was propounded by Descartes and his followers. The Cartesians were trying to identify things that were conserved in physical interactions, and the efforts of Cartesians such as Huygens and Leibniz led to the principles of conservation of momentum and conservation of energy (which they called living force, *vis viva*). Newton, however, proposes here a definition of impressed force that is more closely related to observation and more easily expressed geometrically.

Newton seems to retain something like the older view of impetus in his concept of inherent force (*vis insita*), as presented in Definition 3: both traditional impetus and inherent force remain in the body. Note that he says in his comments on Definition 3 that people are accustomed to think of inertia being in bodies at rest and impetus in moving bodies. Newton replaces both older ideas with his inherent force, saying that motion and rest, "as they are commonly conceived," are distinguished only relative to each other. Impressed force, then, is something new, acting on the body but not remaining with it.

Question for Discussion. If impetus is the same as inertia, it is in the body; it is the body's inherent force. How would you articulate the difference between that force which seems to be in the body (inherent force) and the force which ceases when the impressing action ceases? Is there a transformation of that impressing force from an external action to an internal force? How exactly would you say Newton has altered impetus theory, or has he?

• **Question for Discussion.** Impressed force is called an "action." What does Newton mean by an action? He says in his commentary that impressed force can have its origin in pressure. So apparently this "action" can be operating without any resulting motion if there are contrary forces balancing each other out. If you had to write a definition of "action" that would work for Newton's impressed force, what would you say about it?

• Inherent force and impressed force seem to be the two basic kinds of force. Centripetal force, which will occupy his attention in the next three definitions (and for most of *Principia*), is said here to be an example of impressed force. How that force is impressed in the case of centripetal force is not and cannot be specified in *Principia*. To specify how the centripetal force that is gravity is impressed would be to identify the cause and mechanism of gravity, something Newton does not undertake to do. But as you read and ponder this definition and the Second Law of Motion you might want to keep somewhere in mind that whatever the mechanism of gravity could be, the force of gravity will turn out to be an impressed force under this definition.

• **Question for Discussion.** Have we gotten any more insight into what a force in itself is? How are inherent force and impressed force both sorts of force? How are they different? Is inherent force an "action" in any sense? Presumably in the case of impressed force the action is the action of the impressing force on a body. But is there a sort of negative action in a body's resistance to changing its motion?

Notice also the interesting complementarity in the way Newton defines these two kind of force. The inherent force keeps a body at rest or moving in a straight line. Impressed force is an action that changes a body's state of being at rest or moving in a straight line.

Centripetal Force Defined

Definition 5

Centripetal force is that by which bodies are pulled, pushed, or in any way tend, towards some point from all sides, as to a center.

Of this kind is gravity, by which bodies tend to the center of the earth; magnetic force, by which iron seeks a magnet; and that force, whatever it might be, by which the planets are perpetually drawn back from rectilinear motions and are driven to revolve in curved lines. A stone, whirled around in a sling, attempts to depart from the hand that drives it around, and by its attempt stretches out the sling, doing so more strongly as it revolves more swiftly, and as soon as it is released, it flies away. The force contrary to that attempt, by which the sling perpetually draws the stone back to the hand and retains it in its orbit, I call "centripetal," because it is directed towards the hand as to the center of the orbit. And the account of all bodies that are driven in a gyre is the same. They all attempt to recede from the centers of the orbits, and in the absence of some force contrary to that attempt, by which they are pulled together and kept in their orbits, and which I therefore call "centripetal," they will go off in straight lines with uniform motion.

If a projectile were deprived of the force of gravity, it would not be deflected towards the earth, but would go off in a straight line toward the heavens, doing so with a uniform motion, provided that the resistance of the air be removed. It is drawn back by its gravity from the rectilinear path and is perpetually bent towards the earth, more or less according to its gravity and the velocity of motion. Where its gravity is less in proportion to the quantity of matter, or where the velocity with which it is propelled greater, it will deviate correspondingly less from the rectilinear path, and will travel farther.

If a lead ball, propelled by gunpowder from the summit of some mountain in a horizontal line with a given velocity, were to travel in a curved line for the distance of two miles before it fell to earth, it would travel about twice as far with double the velocity, and about ten times as far with ten times the velocity, provided that the resistance of the air be removed. And by increasing the velocity, the distance to which it is propelled may be increased at will, and the curvature of the line which it describes may be diminished, so that it would finally fall at a distance of ten or thirty or ninety degrees, or it might even go around the whole earth, or, at last, go off towards the heavens, continuing on *in infinitum* with the motion with which it departed. And by the same account, by which a projectile may be deflected into an orbit by the force of gravity and may go around the whole earth, the moon too, whether by the force of gravity (provided it be heavy) or by another force of whatever kind, by which it is urged towards the earth, can be always pulled back towards the earth from its rectilinear path, and deflected into its orbit; and without such a force the moon cannot be held back in its orbit.

This force, if it were less than required, would not sufficiently deflect the moon from the rectilinear path; and if greater than required, would deflect it more than sufficiently, and would lead it down from its orbit towards the earth. It is indeed requisite that it be of exactly the right magnitude, and it is for the Mathematicians to find the force by which a body can be accurately kept back in any given orbit you please with a given velocity, and in turn to find the curvilinear line into which a body departing from any given place you please with a given velocity would be deflected by a given force.

Further, the quantity of this centripetal force is of three kinds: absolute, accelerative, and motive.

Notes on Definition 5

- **Gravity as centripetal force.**

"Of this kind is gravity, by which bodies tend to the center of the earth..."

One perhaps doesn't need a formal proof to recognize that the phenomenon of terrestrial heaviness seems to illustrate a tendency towards a center. If looking around at falling bodies doesn't suggest this, reading Aristotle makes the idea a familiar one.

We must be careful, though. In our day we understand gravity to be more than terrestrial heaviness; we understand it to be the force that moves the heavenly bodies in their curved paths, drawing them out of their tangential motion. If we interpret the statement just quoted from the commentary to this definition as including that too, then we have gotten ahead of ourselves. That assertion must be proved and Newton doesn't assume it until he has proved it. The foundation for the proof is being laid through the first two books and is formally arrived at in Book III Proposition 5.

"And by the same account, by which a projectile may be deflected into an orbit by the force of gravity and may go around the whole earth, the moon too, whether by the force of gravity (provided it be heavy) or by another force of whatever kind, by which it is urged towards the earth, can be always pulled back towards the earth from its rectilinear path, and deflected into its orbit; and without such a force the moon cannot be held back in its orbit."

Notice that Newton is not assuming that the moon is held in orbit by gravity. He says "...whether by the force of gravity (provided it be heavy) or by another force of whatever kind...." Whether the moon is "heavy," in the terrestrial sense of possessing weight, must be shown. Newton intends to prove this; the proof is completed in III.4.

- **Centrifugal force and centripetal force.**

"A stone, whirled around in a sling, attempts to depart from the hand that drives it around, and by its attempt stretches out the sling, doing so more strongly as it revolves more swiftly, and as soon as it is released, it flies away. The force contrary to that attempt, by which the sling perpetually draws the stone back to the hand and retains it in its orbit, I call "centripetal," because it is directed towards the hand as to the center of the orbit."

In the first sentence of this quotation, Newton is alluding to the force that the Cartesians (most notably Huygens) called "*vis centrifuga,*" centrifugal force. This force is a tendency away from the center and is central to their dynamic analysis of all circular motion, including that of the orbits of planets. Newton is shifting our viewpoint here from a supposed tendency away from the center to a tendency towards the center, and has made up the word "centripetal" to refer to that latter tendency.

Absolute Quantity of Centripetal Force Defined

Definition 6

The absolute quantity of centripetal force is the measure of the same, greater or less in proportion to the efficacy of the cause propagating it from the center through the encircling regions.

As the magnetic force is greater in one magnet, less in another, in proportion to the size [*moles*] of the magnet or the intensity of the power [*virtus*].

Notes on Definition 6

- We must again remember that this is a definition of how the term will be used. It is not proof or even assertion that this thing exists. That must be shown. Centripetal force may not be present in the encircling regions around the center; it may be there in the spaces but not propagated from the center. In the last paragraph of the commentary after these last three definitions, Newton will warn that he is considering the forces mathematically, not physically, and that

> ...the reader should beware of thinking that...I am anywhere defining a species or manner of action, or a cause or physical account, or that I am truly and physically attributing forces to centers (which are mathematical points) if I should happen to say either that centers attract, or that forces belong to centers.

• Absolute quantity of centripetal force is not mentioned again until I.17, and then it is, as always prior to Book III, hypothetical. That is, I.17 says that if we suppose there to be given a certain absolute quantity of force, along with other conditions, he will show us how to determine a resulting curve.

• Notice again that we are given not an equation but a proportion. See the note to Definition 1.

• **Question for Discussion.** Does the idea of force propagated out into surrounding spaces sound occult to you? If it doesn't, should it? How could this absolute quantity of force be measured? How would you propose it be measured?

Keep these speculations in mind as you read the following definitions.

Accelerative Quantity of Centripetal Force Defined

Definition 7

The accelerative quantity of centripetal force is the measure of the same, proportional to the velocity which it generates in a given time.

Thus the power of the same magnet is greater at a less distance, less at a greater one; or the gravitating force is greater in valleys, less on the peaks of high mountains, and less still (as will become clear hereafter) at greater distances from the earth's globe. At equal distances, however, it is everywhere the same, because it equally accelerates all falling bodies (heavy or light, great or small) once the resistance of the air is removed.

Notes on Definition 7

• We again have a proportion. We are instructed that we are to compare two centripetal forces by comparing the velocity each generates in a given time. We are not being told how to calculate the magnitude or number of a particular force or of its accelerative quantity.

• Note that neither Definition 7 nor Definition 8 assumes or defines an instantaneous force. He says "in a given time." This time could be taken to the limit (gradually reduced towards zero), thus yielding a rato of forces at particular points, but it could also be finite.

Look out for assuming that accelerative quantity of force must be the second derivative of distance. Aside from the fact that this definition gives us, as noted in the previous bullet, a ratio of forces and not a value for a particular force, there is another problem with making that conscious or unconscious translation (a translation that can and has led to serious misunderstandings of some of his proofs even by scholars who should know better). Modern physics has become very comfortable, even blasé, about operating on infinitesimals, but Newton is very careful in his work with vanishing quantities. He doesn't throw around ratios of infinitesimals, but rather takes ratios of finite magnitudes and looks at what happens to those ratios as time, or as one of the magnitudes, shrinks towards zero.

In Newton's first set of Lemmas (not included in this selection), he demonstrates his mastery in establishing and maintaining a firm grip on the notoriously slippery creatures that vanishing quantities are.

• Be careful about Newton's commentary. The italicized part is his actual definition. The commentary following it is looking ahead at what he will develop formally. His readers know that experiments indicated the empirical truth of these claims about the magnet and the power of the gravitational force on the earth's surface, but note that the definition says nothing about the force varying as the distance. As you begin Newton's exposition in the next section, Section 2, "On the Finding of Centripetal Forces" (it begins with Proposition 1), you will be astonished by how little he assumes. All this discussion in these commentaries to the Definitions and Laws, and the scholia after the Definitions and Laws, will be ignored. That gravitational force varies with the distance will be brought out with the greatest care, and not established until Book III. Nor will he assume the equal acceleration of light and heavy bodies at equal distances without a formal demonstration; this too is established finally in Book III.

• If you are thinking that accelerative quantity of force, measured only by the change in velocity it effects, is not the "proper" definition of force, you are no doubt being overly influenced by what you learned in your physics class. Set that aside.

Let Newton unfold his system to you. He has good reasons for what he does. If you think you want to be dealing with mass right from the start, you don't know what you would be getting yourself in for. When Newton finally brings mass in, in Book III, first in the attracted body and then in the attracting body, he will have carefully built up a very complex system adding one layer of complexity at a time.

• **Question for Discussion.** We asked in thinking about the previous definition how we might measure the absolute quantity of force, "the efficacy of the cause propagating it from the center through the encircling regions." Would this definition offer us a way to measure it? Newton is suggesting that we look at how much velocity is generated in a given time. If we look at those velocities in different places, perhaps we could say that we were measuring the efficacy of the force at those places. What do you think?

Motive Quantity of Centripetal Force Defined

Definition 8

The motive quantity of centripetal force is the measure of the same proportional to the motion which it generates in a given time.

Thus the weight is greater in a greater body, less in a lesser one, and in the same body it is greater near the earth, less in the heavens. This quantity is the whole body's centripetency, or propensity towards the center, and (if you will) weight, and is always known through the force contrary and equal to it, by which the descent of the body can be prevented.

Notes on Definition 8

• We again have a proportion. The definition says that motive forces are proportional to their respective motions generated in a given time.

A shorthand way of saying this is "motive force varies as motion generated in a given time." But if you do use this shorthand, it is important to understand that, if you're talking about Newton, your shorthand formulation means comparing two (or more) motive forces by comparing the amount of motion (defined in Definition 2) each has generated in a particular time segment.

We are instructed in the way to compare two centripetal forces. We are not being told how to calculate the magnitude or number of a particular force or of its motive quantity.

• **"a given time."** This is not a definition of an instantaneous force; see the second bulleted note under Definition 7.

• **Weight.** The argument that weight is an example of motive force comes in the paragraphs that follow in Newton's text, in his commentary on Definitions 6–8.

- **Motive Force in Contrast to Accelerative Force.** Recall the definition of accelerative quantity of force (Definition 7). Those forces are proportional to the respective velocities generated in a given time. In contrast, the motive quantities of force are proportional to the respective quantities of motion (Definition 2) generated in a given time. The difference here is the introduction of the mass of the attracted body.

The mass of the attracted body (the body on which the centripetal force is exerted) will not become relevant to the development of universal gravitation until Book III Proposition 6; in fact it won't even be mentioned in Newton's central argument until then. He must lay much groundwork just looking at accelerative force before he can bring in that next level of complication.

- The reader may have recognized motive force as the sort of force modern physics has in mind when it uses the equation $f = ma$. We've reminded ourselves in the first note above that Newton doesn't give us an equation here, but rather a proportion, relationships between instances of like things.

Newton's Commentary to Definitions 6–8

For the sake of brevity, these quantities of force may respectively be called motive forces, accelerative forces, and absolute forces, and for the sake of distinction, may be said to trace their origins respectively to the bodies seeking the center, to the places of the bodies, and to the center of forces. That is, motive force traces its origin to the body, as if the force were the endeavor of the whole toward the center, composed of the endeavors of all the parts; accelerative force traces its origin to the place of the body, as if it were a sort of efficacy, spread out from the center through the individual places on the circumference, for moving bodies that are in those places; and absolute force traces its origin to the center, as if it were endowed with some cause without which the motive forces would not be propagated through the regions on the circumference, whether that cause be some central body (such as is a magnet at the center of the magnetic force, or the earth in the center of the gravitating force) or something else that does not appear. This concept is strictly mathematical, for I am not now considering the causes and physical seats of the forces.

The accelerative force is accordingly to the motive force as speed is to motion. For the quantity of motion arises from the speed and the quantity of matter, and the motive force arises from the accelerative force and the quantity of the same matter conjoined. For the total action of the accelerative force on the individual particles of a body is the motive force of the whole. Whence, near the surface of the earth, where the accelerative gravity or the gravitating force is the same in all bodies, the motive gravity, or weight, is as the body; but if an ascent be made to regions where the accelerative gravity is less, the weight gradually decreases, and will always be as the body and the accelerative gravity conjoined. Thus in regions where the accelerative gravity is less by a factor of two, the weight of a body that is smaller by a factor of two or three will be less by a factor of four or six.

Further, I call attractions and impulses accelerative and motive in the same sense. Moreover, I use the words "attraction," "impulse," or [words denoting] a propensity of any kind toward a center, indifferently and promiscuously for each other: I am considering these forces, not physically, but only mathematically. Therefore, the reader should beware of thinking that by words of this kind I am anywhere defining a species or manner of action, or a cause or physical account, or that I am truly and physically attributing forces to centers (which are mathematical points) if I should happen to say either that centers attract, or that forces belong to centers.

Notes on Commentary to Definitions 6–8

• **Disclaimers on Cause of Gravity.** Pay special attention to the whole last paragraph above, as well as to Newton's earlier statement, "this concept is strictly mathematical, for I am not now considering the causes and physical seats of the forces," which also disavows any claim that the power comes from the center or a central body.

A reader of *Principia* needs to keep these Newtonian warnings in mind, because Newton will indeed talk about centers and attractions and impulses, and one could indeed get a distorted idea, not only of his physical picture but the structure of his argument, if one were to think he was attributing physical reality to things that are not seen so by him. It is a vulgar modern idea that bodies act on other bodies at a distance; Newton found the idea absurd.

None of this, however, prevents one from deeply pondering, in working through *Principia*, what the cause of gravity might be, and to ponder as well what Newton's speculations might have been.

Scholium on absolute time, space, place, and motion

Scholium

Hitherto it has seemed appropriate to explain the less familiar terms, [and] the sense in which they are to be taken in what follows. Time, space, place, and motion, are very familiar to everyone. It should nevertheless be noted that these are not commonly conceived of otherwise than from their relation to sensible objects. And from this there arise certain prejudices, for the removal of which it is useful for these same [terms] to be distinguished into absolute and relative, true and apparent, mathematical and common.

I. Absolute, true, and mathematical time, in itself and by its nature without relation to anything external, flows uniformly, and by another name is called "duration." Relative, apparent, and common [time] is the perceptible and external measure (whether accurate or varying in rate) of any duration you please by means of motion, which is commonly used in place of true time, such as an hour, a day, a month, a year.

II. Absolute space, by its nature, without relation to anything external, always remains similar and motionless. Relative [space] is any movable measure or dimension you please of this space, which [measure] is defined by our senses through its position with respect to bodies, and is commonly taken in place of motionless space, such as the dimension of subterranean, aerial, or celestial space defined through its position with respect to earth. Absolute and relative space are the same in form and size, but do not always remain the same in number. For if the earth, for example, were to move, the space of our air, which relatively and with respect to our earth always stays the same, will be now one part of absolute space in which the air moves across, now another part of it, and thus, absolutely, it will perpetually change.

III. Place is the part of space which a body occupies, and is absolute or relative according to the space. It is a part, I say, of space, not the location of a body, or the enclosing surface. For the places of equal solids are always equal; the surfaces, however, are nearly always unequal because of the dissimilarity of figures. Locations, on the other hand, do not, properly speaking, have quantity, nor are they so much places as properties of places. The motion of the whole is the same as the sum of the motions of the parts: that is, the translation of the whole from its place is the same as the sum of the translations of the parts from their places. Consequently, the place of the whole is the same as the sum of the places of the parts, and for that reason it is internal and in the whole body.

IV. Absolute motion is the translation of a body from absolute place to absolute place; relative [motion is the translation of a body] from relative [place] to relative [place]. Thus in a boat which is carried with sails set, the relative place of a body is that region of the boat in which the body is, or that part of the whole concavity which the body fills, and which to that extent moves along with the boat; and relative rest is the body's continuing to remain in the same region of the boat or part of the concavity. But true rest is the body's continuing to remain in the same motionless part of that space in which the boat itself along with its concavity and all its contents moves. Whence if the earth is really at rest, a body which is relatively at rest in the boat will move truly and absolutely with that velocity with which the boat moves upon the earth. If on the contrary the earth also moves, the true and absolute motion of the body arises partly from the true motion of the earth in motionless space, partly from the relative motion of the boat upon the earth; and if the body also moves relatively in the boat, its true motion arises partly from the true motion of the earth in motionless space, partly from the relative motions of the boat upon earth and of the body in the boat; and from these relative motions there arises the body's relative motion upon earth. Thus, if that part of earth where the boat is placed be really moved eastward with a velocity of 10,010 units, and the boat be carried by the sails and the wind westward with a velocity of ten units, while the boatman walk on the boat towards the east with a velocity of one unit, the boatman will move truly and absolutely in motionless space with 10,001 units of velocity eastward, and relatively upon earth towards the west with nine units of velocity.

Absolute time is distinguished from relative in astronomy by the equation of common time. For the natural days are unequal, but are commonly taken as if equal for a measure of time. Astronomers make a correction for this inequality, so that they may measure the celestial motions from a truer time. It is possible that there is no uniform motion by which time may be measured accurately. All motions can be accelerated and retarded, but the flow of absolute time cannot be changed. The duration or perseverance of the existence of things is the same, whether the motions are fast or slow or none. Furthermore, these are rightly distinguished from their perceptible measures, and are reckoned from them through the astronomical equation. Moreover, the need for this equation in determining phenomena is established both by the experimental evidence of the pendulum clock and by the eclipses of Jupiter's moons.

As the order of the parts of time is unchangeable, so also is the order of the parts of space. If these move from their places, they will also (so to speak) move from themselves. For times and spaces are in a way the places of themselves and of all things. Everything without exception is located in time according to order of succession, in space according to order of place. It is of their essence that they be places, and it is absurd for primary places to move. These are therefore absolute places, and translations from these places are alone absolute motions.

However, because these parts of space cannot be seen and cannot be distinguished from each other by our senses, we introduce perceptible measures in their stead. For from the positions and distances of things from some body, which we see as motionless, we define all places universally, and thereafter we also estimate all motions as well with respect to the places mentioned previously, insofar as we conceive the bodies to be carried away from them. We thus use relative places and motions in place of absolute ones, and this is not an inconvenience in human affairs. In philosophical matters, however, an abstraction from the senses must be made. For it can happen that there is no body really at rest, to which places and motions may be referred.

Further, absolute and relative motions are distinguished from each other through their properties and causes and effects. It is a property of rest that bodies really at rest are at rest among themselves. Therefore, since it is possible that some body in the regions of the fixed stars, or far beyond, be absolutely at rest, while it cannot be known from the position of bodies with respect to each other in our regions whether one of them may preserve its given position with respect to that distant [body], true rest cannot be defined from the position of the latter [bodies] among themselves.

It is a property of motion that the parts, which preserve given positions with respect to the wholes, participate in the motions of the same wholes. For all parts of bodies moving in curves strive to recede from the axis of motion, and the impetus of bodies moving forward arises from the conjoined impetus of the individual parts. Therefore, when the surrounding bodies are moving, those are moving which are at rest among the surrounding ones. And for that reason, true and absolute motion cannot be defined through translation away from nearby bodies, which are viewed as if they were resting bodies. For the external bodies ought not only to be viewed as if they were resting, but also to be truly at rest. Otherwise, all the surrounded [motions], other than translation away from the nearby surrounding [bodies], will also participate in the true motions of the surrounding [bodies], and when that translation is removed, they are not truly at rest but will only be viewed as if they were resting. For the surrounding are to the surrounded as the exterior part of the whole is to the interior part, or as the shell to the kernel. And when the shell is moved, the kernel too is moved without translation away from the surrounding shell, as a part of a whole.

Related to the preceding property is that when the place is moved the thing in the place is moved along with it: thus a body that is moved away from a moved place, also participates in the motion of its place. Therefore, all motions which take place away from [i.e., with respect to] moving places are only parts of the whole and absolute motions, and every whole motion is compounded of the motion of a body from its prime place, and the motion of that place from its place, and so on, until it comes to a motionless place, as in the example of the boatman mentioned above. Hence, whole and absolute motions can be defined only through motionless places, and I have consequently related them above to motionless places, and relative [motions] to movable [places]. However, only those places are motionless which from infinity to infinity all preserve given positions with respect to each other, and moreover remain forever motionless, and constitute that space which I call immovable.

The causes by which true and relative motions are distinguished from each other are the forces impressed upon bodies for generating motions. True motion is neither generated nor changed except by forces impressed upon the moved body itself. But relative motion can be generated and changed without forces being impressed upon this body. For it suffices that they be impressed only upon other bodies to which it is related, so that, when they give way, that relation, of which the relative rest or motion of this [body] consists, is changed. Again, true motion is always changed by forces impressed upon the moved body, but relative motion is not necessarily changed by these forces. For if the same forces were impressed upon other bodies as well, to which there is a relation, in such a way that the relative position be preserved, the relation in which the relative motion consists will be preserved. Therefore, every relative motion can be changed while the true motion is preserved, and can be preserved while the true motion is changed, and for that reason true motion consists not at all in relations of this sort.

The effects by which absolute and relative motions are distinguished from each other are the forces of receding from the axis of circular motion. For in purely relative circular motion, these forces are none, but in true and absolute [circular motion] they are greater or less according to the quantity of motion.

Suppose that a pail should hang from a very long cord, and be driven continually in a circular path, until the cord becomes somewhat stiff from twisting, and [the pail] next be filled with water, and be at rest along with the water, and then be driven by some sudden force in a circular path with a contrary motion, persevering in that motion for a long time as the cord untwists. At the beginning, the surface of the water will be flat, as it was before the motion of the vessel, but after the vessel, by a force gradually impressed upon the water, makes it too begin to rotate perceptibly, it will itself gradually withdraw from the middle, and will climb up to the sides of the vessel, adopting a concave form (as I have myself experienced); and, with an ever increasing motion, will climb more and more, until it make its revolutions in an equal time with the vessel, and come to rest relative to it. This climbing is an indicator of a striving to withdraw from the axis of motion, and through such a striving the true and absolute circular motion of the water, here completely contrary to relative motion, comes to be known and is measured. At first, where the relative motion of the water in the vessel was greatest, that motion did not arouse any striving to withdraw from the axis: the water was not seeking the circumference by climbing up the sides of the vessel, but stayed flat, and for that reason its true circular motion had not yet begun. Later, however, where the relative motion of the water decreased, its ascent up the sides of the vessel was an indicator of the striving to withdraw from the axis, and this striving showed its true circular motion, which was ever increasing, and was finally made greatest where the water came to rest relatively in the vessel. Therefore, this striving will not depend upon the translational motion of the water with respect to the surrounding bodies, and consequently, true circular motion cannot be defined by such translations.

The true circular motion of any revolving [body] is unique, corresponding to a unique striving which is its own, as it were, and is commensurate with the effect. Relative motions, however, are countless, in accord with the various relations with external [bodies], and, like relations, they are entirely bereft of true effects, except insofar as they participate in that true and unique motion. Hence also, in the system of those people who would have our heavens rotate in an orb beneath the heavens of the fixed stars, and bear the planets along with them, the individual parts of the heavens, and the planets which are indeed relatively at rest in the heavens nearest them, in truth move. For they change their positions with respect to each other (unlike what happens in those truly at rest), and, being carried along with the heavens, they participate in their motion, and, like the parts of all revolving things, strive to recede from their axes.

Relative quantities are therefore not those quantities themselves whose names they display, but are those perceptible measures of them (whether true or erroneous) which are commonly used in place of the measured quantities. And if the meanings of words should be defined from usage, then by those names "time," "space," "place," and "motion," these perceptible measures should properly be understood, and the discussion will be out of the ordinary and purely mathematical if the quantities measured be understood here. Furthermore, they do violence to sacred scripture who interpret these words as concerning measured quantities there. Nor are mathematics and philosophy any the less defiled by those who confuse the true quantities with their relationships and common measures.

To recognize the true motions of individual bodies and to distinguish them in fact from the apparent ones, is indeed extremely difficult, for the reason that the parts of that motionless space in which the bodies truly move do not flow in to the senses. Nevertheless, the cause is not entirely hopeless. For arguments can be taken, partly from the apparent motions which are the differences of true motions, partly from the forces which are the causes and effects of the true motions.

As, if two globes, joined together at a given distance from each other by a cord between them, were to revolve about their common center of gravity, the striving of the globes to recede from the axis of motion would come to be known from the tension of the cord, and from that the quantity of circular motion might be computed. Then if any equal forces you please were to be simultaneously impressed upon alternate faces of the globes, to increase or decrease the circular motion, the increase or decrease of the motion would come to be known from the increased or decreased tension of the cord, and finally, it would be possible from that to find the faces of the globes upon which forces should be impressed in order most greatly to increase the motion, that is, the aftermost faces, or those which are the following ones in the circular motion. And when the faces which are following are known, and the opposite faces which are leading, the direction of motion would be known. In this way, both the quantity and the direction of this circular motion might be found in whatever immense void you please, where nothing external and perceptible were to exist with which the globes might be compared. Now, if there were to be set up in that space some distant bodies maintaining a given position among themselves, such as are the fixed stars in the regions of the heavens, it would indeed be impossible to tell from the relative translation of the globes within the bodies whether the motion should be attributed to the former or the latter. But if attention were paid to the cord, and it were ascertained that its tension were exactly that which the motion of the globes would require, it would be permissible to conclude that the motion belonged to the globes, and that the bodies were at rest, and, finally, from the translation of the globes within the bodies, to determine the direction of the motion. But how to determine the true motions from their causes, effects, and apparent differences, and, conversely, how to determine their causes and effects from the motions, whether true or apparent, will be taught more fully in what follows. For it is to this end that I wrote the following treatise.

Axioms, or Laws of Motion

Law 1

Every body continues in its state of resting or of moving uniformly in a straight line, except insofar as it is driven by impressed forces to alter its state.

Projectiles continue in their motions except insofar as they are slowed by the resistance of the air, and insofar as they are driven downward by the force of gravity. A top, whose parts, by cohering, perpetually draw themselves back from rectilinear motions, does not stop rotating, except insofar as it is slowed by the air. And the greater bodies of the planets and comets preserve their motions, both progressive and circular, carried out in spaces of less resistance, for a longer time.

Note on Law 1

● This law in effect asserts the existence of *vis insita,* inherent force or inertia, defined in Definition 3. Now it can be used, as it is in Proposition 1.

Law 2

The change of motion is proportional to the motive force impressed, and takes place following the straight line in which that force is impressed.

If some force should generate any motion you please, a double [force] will generate a double [motion], and a triple [force] a triple [motion], whether it has been impressed all at once, or gradually and successively. And because this motion is always directed in the same way as the generating force, if the body was previously in motion, then this [impressed] motion is either added to its motion (if they have the same sense) or subtracted (if contrary), or joined on obliquely (if oblique) and compounded with it according to the determination of the two.

Notes on Law 2

● See Definition 4. More is added to the definition of impressed force through this law. We are now told that impressed force is a motive force.

Note that we are told this about impressed forces generally, and by Definition 4 we know that impressed forces can be of various kinds, only one of which is centripetal force. When we read "motive force" here, therefore, the words are not elliptical for "motive quantity of centripetal force" (Definition 8). Both might have the same measure (both being proportional to quantity of motion), but motive force may cover forces which are not centripetal.

- "...whether it has been impressed all at once, or gradually and successively."

Motive force is defined as taking place "in a given time." Thus the impressed force whose action is the subject of this law takes place as Newton describes in his commentary (quoted above). We note that Newton does not envision impressed force as an instantaneous action.

Law 3

To an action there is always a contrary and equal reaction; or, the mutual actions of two bodies upon each other are always equal and directed to contrary parts.

Whatever pushes or pulls something else is pushed or pulled by it to the same degree. If one pushes a stone with a finger, his finger is also pushed by the stone. If a horse pulls a stone tied to a rope, the horse will also be equally pulled (so to speak) to the stone; for the rope, being stretched in both directions, will by the same attempt to slacken itself urge the horse towards the stone, and the stone towards the horse, and will impede the progress of the one to the same degree that it promotes the progress of the other. If some body, striking upon another body, should change the latter's motion in any way by its own force, the same body (because of the equality of the mutual pushing) will also in turn undergo the same change in its own motion, in the contrary direction, by the force of the other. These actions produce equal changes, not of velocities, but of motions — that is, in bodies that are unhindered in any other way. For changes in velocities made thus in opposite directions, are inversely proportional to the bodies, because the motions are equally changed. This law applies to attractions as well, as will be proven in the next Scholium.

Note on Law 3

- Newton, in the last sentence above, extends the workings of this law to apply to "attractions" as well as conventional mechanical pushes and pulls. He says that this will be proved in the scholium which follows the corollaries.

This final sentence in the commentary to Law 3 was added in the second edition of *Principia*, along with the two thought experiments in the scholium that constitute the promised proof. Consider Law 1 and Corollary 4 as you evaluate these thought experiments. After considering what Newton says, you may wish to read my note which discusses this extension of Law 3. That note follows the scholium.

• Of course, a law of motion is a law, not a proposition. It doesn't need to be proved. It is like a postulate, but not a hypothetical postulate of the form "suppose we assume this, then what would follow?" It resembles more a statement of something sufficiently clear from experience as not to need a proof. It may be a new formulation, a new way of thinking about things, or even a new observation; but it is asserted with no fear that contradictory instances will be found.

This is clearly the case with the earlier parts of the statement of Law 3, the ones involving pushes and pulls of a direct mechanical sort. The case with attractions has a somewhat different status. We don't have hands-on experience of the workings of third law interactions in the case of attraction.

We may even have the sense of attractions being independent of third law reactions or reciprocation, as they tend to be for the attractions arising within the souls of living beings. (By soul I mean that element in living beings—sometimes called psyche or consciousness or even "heart"—that makes them more than inert matter.)

And we start to wonder what "attractions" between lumps of *matter* could be. Since they don't have souls, it must be some push or pull mechanism, but what? Depending on the mechanism, the third law might or might not apply.

For this reason Newton feels that he needs to offer a proof specifically for attractions. He says it will be proved (Latin *probabitur*) in the scholium, but what he offers is more like a reassurance that our universe does indeed work as if gravitational attractions were obeying the third law.

Corollaries to the Laws of Motion

Note on Corollaries to the Laws

The note after the scholium following these corollaries will raise a question about the relationship between Law 3 and Corollary 4, so you will want to be alert to what emerges about that relationship as you read the corollary.

Newton presents three laws and then gives us six corollaries. But to *what*, exactly, are they corollaries? One might wonder whether they are all corollaries to Law 3, all corollaries to all the laws, or connected to the laws in various other ways.

Corollary 1

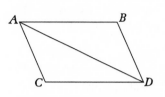

A body [urged] by forces joined together, describes the diagonal of a parallelogram in the same time in which it describes the sides separately.

Suppose that, in a given time, by the force M alone, impressed at place A, a body would be carried with uniform motion from A to B, and by the force N alone, impressed at the same place, it would be carried from A to C. Let the parallelogram $ABCD$ be completed, and that body will be carried by both forces in the same time on the diagonal from A to D. For because the force N acts along the line AC parallel to BD, this force, by Law 2, will not at all change the velocity of approach to that line BD generated by the other force. Therefore, the body will arrive at the line BD in the same time whether force N be impressed or not; and therefore, at the end of that time, it will be found somewhere on that line BD. By the same argument, at the end of the same time, it will be found somewhere on the line CD, and for that reason it must necessarily be found at the intersection of the two lines D. But, by Law 1, it will proceed with a rectilinear motion from A to D.

Corollary 2

And hence is evident the composition of a direct force AD from any oblique forces you please AB and BD, and, in turn, the resolution of any force you please AD into any oblique ones whatever AB and BD. This composition and resolution is, moreover, abundantly confirmed from mechanics.

[Newton does not offer a proof of Corollary 2. Instead, he gives an example of its application to "wheels, winches, pulleys, levers, taut cords, and weights"—the basic elements out of which, he asserts, all machines are compounded. In this way he shows that the principle of composition and resolution of forces pervades the whole doctrine of mechanics.

As his example he describes a delightfully elaborate contraption that has come to be known as "Newton's wheel," though it involves far more than a wheel. But the analysis is long and complicated and so is omitted here.]

Corollary 3

The quantity of motion that is obtained by taking the sum of the motions made in the same direction, and the difference of those made in opposite directions, is not changed by action of the bodies among themselves.

[The proof of this corollary is omitted here.]

Corollary 4

The common center of gravity of two or more bodies does not change its state either of motion or of rest by actions of the bodies among themselves, and for that reason the common center of gravity of all bodies acting mutually upon one another (external actions and hindrances being excluded) either is at rest or moves uniformly in a straight line.

[The proof of this corollary is omitted here.]

Corollary 5

The motions of bodies contained in a given space are the same among themselves, whether that space be at rest, or whether it move uniformly in a straight line without circular motion.

For the differences of motions tending in the same direction, and sums of those tending in opposite directions, are the same at the beginning in both cases (by supposition), and from these sums or difference arise the collisions and impulses with which the bodies strike each other. Therefore, by Law 2, the effects of the collisions will be equal in both cases, and therefore the motions among themselves in the one case will remain equal to the motions among themselves in the other. The same is confirmed by a most lucid experiment. All motions on a ship relate to each other in the same way whether the ship be at rest or move uniformly in a straight line.

Corollary 6

If bodies be moved in any manner whatever among themselves, and be urged by equal accelerative forces along parallel lines, everything goes on moving in the same manner among themselves as if they had not been impelled by those forces.

For those forces, in acting equally (in proportion to the quantities of the bodies to be moved) and along parallel lines, will move all the bodies equally (as regards velocity), by Law 2. Therefore, their positions and motions among themselves will not change.

Scholium After the Laws of Motion

Scholium

Up to this point, I have presented principles accepted by mathematicians and confirmed by manifold experience. By the first two laws and the first two corollaries, Galileo found that the descent of heavy bodies is in the duplicate ratio of the time, and that the motion of projectiles takes place in a parabola, in agreement with experience, except to the extent that those motions are retarded by some small amount by the resistance of air. When a body falls, uniform gravity, acting equally in the individual equal particles of time, impresses equal forces upon that body, and generates equal velocities; and in the whole time impresses a whole force and generates a whole velocity proportional to the time. And spaces described in proportional times are as the velocities and the times conjointly, that is, in the duplicate ratio of the times. And when a body is projected upwards, uniform gravity impresses forces and subtracts velocities proportional to the times, and the times of ascending to the highest altitudes are as the velocities to be subtracted, and those altitudes are as the velocities and the times conjointly, or in the duplicate ratio of the velocities. And the motion of a body projected along any straight line you please, arising from the projection, is compounded with the motion arising from gravity.

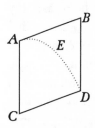

As, for example, a body A might, by the motion of projection alone, describe the straight line AB in a given time, and by the motion of falling alone might in the same time describe the height AC. Let the parallelogram $ABDC$ be completed, and at the end of the time that body in its compound motion will be found at place D, and the curved line AED, which that body will describe, will be a parabola which the straight line AB touches at A, and whose ordinate BD is as sq. AB. From the same laws and corollaries depend matters demonstrated of the times of swinging pendulums, with the support of everyday experience of clocks.

From these same things and the Third Law, Sir Christopher Wren, Dr. John Wallis, and Christian Huygens, easily the chief geometers of recent times, have found the rules of collisions and rebounds of two bodies, and communicated them to the Royal Society at nearly the same time, fully in agreement with each other (as regards these laws); and in fact first Wallis, then Wren and Huygens, produced the discovery. But the truth was also confirmed by Wren before the Royal Society by an experiment of pendulums, which the most illustrious Mariotte also soon saw fit to expound in an entire book. In fact, to make this experiment match the theories exactly, one must take into account the resistance of air, as well as the elastic force of the colliding bodies.

...

In attractions, I show the matter briefly thus. There being any two bodies you please A, B, that mutually pull each other, conceive of any obstacle you please interposed, by which their coming together may be impeded. If one of the bodies A is pulled more towards the other body B than that other B [is pulled] towards the former A, the obstacle will be urged more by the pressure of the body A than by the pressure of the body B, and accordingly will not remain in equilibrium. The stronger pressure will prevail, and will make the system of the two bodies and the obstacle move in a straight line towards the parts in the direction of B, and go off *in infinitum* in free spaces with a motion always accelerated. Which is absurd and contrary to the First Law. For by the First Law the system will be obligated to persist in its state whether of resting or of moving uniformly in a straight line, and accordingly the bodies will urge the obstacle equally, and for that reason will be equally pulled towards each other. I have tried this on a magnet and iron. If these two be placed by themselves in separate vessels that touch each other, in still water they will float together; neither will propel the other, but by an equality of attraction on both sides they will maintain equal strivings towards each other, and at length, being set in equilibrium, will come to rest .

Thus also, gravity between the earth and its parts is mutual. Let the earth FI be cut by any plane you please EG into two parts EGF and EGI, and the weights of these towards each other mutually will be equal. For if by another plane HK, which is parallel to the former EG, the greater part EGI be cut into two parts $EGKH$ and HKI, of which HKI is equal to the former part cut off EFG, it is manifest that

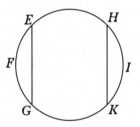

the middle part $EGKH$ will by its own weight be inclined towards neither of the extreme parts, but will be suspended between the two in equilibrium, so to speak, and will be at rest. But the extreme part HKI will press on the middle part with its whole weight, and will urge it towards the other extreme part EFG, and therefore the force by which the sum EGI of the parts HKI and $EGKH$ tends towards the third part EGF is equal to the weight of the part HKI, that is, to the weight of the third part EGF. And on this account the weights of the two parts EGI, EGF to each other mutually are equal, just as I wished to show. And unless those weights were equal, the whole earth, floating in the free ether, would give way to the greater weight, and in fleeing from it would go off *in infinitum*.

As bodies have the same strength in collision and rebound, whose velocities are inversely as the inherent forces, likewise, in making mechanical instruments move, agents have the same power and support each other by contrary efforts, whose velocities, reckoned along the direction of the forces, are inversely as the forces.

Thus, weights will be equivalent in moving the arms of a balance, which, when the balance swings, are inversely as their velocities upwards and downwards; that is, weights, if they ascend and descend directly, will be equivalent, which are inversely as the distance from the balance's axis of the points from which they are suspended. If, however, they ascend or descend obliquely, impeded by inclined planes or other obstacles set in the way, [those weights] will be equivalent, which are inversely as the ascents and descents, insofar as they are made along the perpendicular: this is because gravity is directed downward.

Similarly, in a winch or block and tackle, the force of the hand directly pulling the rope, which is to the weight, ascending either directly or obliquely, as the velocity of perpendicular ascent is to the velocity of the hand pulling the rope, will support the weight.

In clocks and similar instruments, which are constructed of linked wheels, the contrary forces for furthering and hindering the motion of the wheels, if they are inversely as the velocities of the parts of the wheels upon which they are impressed, will support each other.

The force of a screw for pressing a body is to the force of the hand driving the handle around, as the circular velocity of the handle at that part where the hand urges it, is to the forward velocity of the screw towards the pressed body. The forces by which a wedge urges the two parts of a split timber are to the force of the hammer on the wedge, as the forward motion of the wedge along the direction of the force impressed upon it by the hammer, is to the velocity by which the parts of the timber give way to the wedge, along lines perpendicular to the faces of the wedge. And the account of all machines is comparable.

The efficacy and usefulness of these consists in this alone, that in decreasing the velocity we may increase the force, and vice versa. From this comes the solution, in every kind of suitable device, of the problem, "to move a given weight with a given force," or otherwise, to overcome a given resistance with a given force. For if machines be so arranged that the velocities of the agent and the resistance are inversely as the forces, the agent will support the resistance, and with a greater disparity of velocities, will overcome the same. Certainly, if the disparity of velocities be so great that all resistance is overcome, that normally arises from the wearing away of bodies that are contiguous and rubbing each other, as well as from the cohesion of [bodies that are] continuous and that must be separated from each other, and [from] the weights of [bodies that are] to be lifted, once all resistance is overcome, the force left over will produce an acceleration of motion proportional to itself, partly in the parts of the machine, partly in the resisting body. Nevertheless, to treat of mechanics is not the purpose of this. In these remarks, I have intended only to show how broadly evident and how certain the Third Law of Motion is. For if the action of an agent is reckoned from its force and velocity conjointly, and the reaction of the resistance is likewise reckoned conjointly from the velocities of its individual parts and the forces of resistance arising from their wearing away, cohesion, weight, and acceleration, the action and reaction, in every use of instruments, will always be equal to each other. And insofar as an action is propagated through the instrument and is ultimately impressed upon every resisting body, its final direction will always be contrary to the direction of the reaction.

Note on Law 3 and Scholium to the Laws

The following notes raise a few questions about Newton's application of Law 3 to gravitational attractions as supported by the two thought experiments in the preceding scholium. You might want to reread or refer back to Law 3 and its commentary (including the last sentence in which he says that that he proves in the scholium that Law 3 applies to attractions as well as mechanical pushes and pulls) and to the thought experiments of the scholium on pages 35 and 36. The two thought experiments are given in the two paragraphs beginning with the words "In attractions, I show the matter briefly thus."

• **What are "attractions"?** I put quotation marks around attractions, because it is not clear exactly what attractions are when we are talking about the physical world and not about movements of the soul. The thought experiments involve gravitational attractions, so perhaps we can at least narrow the focus down somewhat. But we remember Newton's emphasis in his commentary after the Definitions that when he speaks of attraction he only means it mathematically, not physically, and says "I use the words 'attraction,' 'impulse,' or words denoting a propensity of any kind toward a center, indifferently and promiscuously for each other… Therefore, the reader should beware of thinking that by words of this kind I am anywhere defining a species or manner of action, or a cause or physical account."

This means that the physical cause or mechanism for what he calls attraction may be impulse(s) from behind the attracted body—differential ether pressure, for example. If so, the physical action would not be between the two bodies: there would indeed be Law 3 interactions but they would be between each body and the ether particles impelling it.

• **Questions for Discussion:** What do you make of Newton's two thought experiments showing that Law 3 applies to "attractions"? Has the assertion made in the last sentence of his commentary to Law 3 that the law applies been, as he said, "proven"? And, considering that this is a law and not a proposition, is proof even the right term? Laws are laws of nature. They are like postulates. And isn't Newton, in his exposition, in fact proceeding as if they are postulates? One might at least get the impression that he is calling on our experience of the world in observing that the earth, or the system of two bodies with an obstacle between them, remains at rest and doesn't go accelerating out in some direction to infinity.

And yet, what about his formal invoking of Law 1? He says that the bodies don't fly off because that would violate Law 1. So what is really at issue here, Law 3 or Law 1? And how exactly does the acceleration of the two bodies, in the case he declares absurd, violate Law 1? Law 1 only applies to bodies in the absence of an impressed force. But gravitational attraction, centripetal force, is an impressed force, as Newton tells us in Definition 4.

SECTION 2

On the Finding of Centripetal Forces

Proposition 1

The areas which bodies driven in orbits [gyros] *describe by radii drawn to an immobile center of forces, are contained in immobile planes and are proportional to the times.*

Let the time be divided into equal parts, and in the first part of the time let the body, by its inherent force, describe the straight line AB. In the second part of the time, the same body, if nothing were to impede it, would pass on by means of a straight line to c (by Law 1), describing the line Bc equal to AB, with the result that, radii AS, BS, cS being drawn to the center, the areas ASB, BSc would come out equal. But when the body comes to B, let the centripetal force act with an impulse that is single but great, and let it have the effect of making the body depart from the straight line Bc and continue in the straight line BC. Let

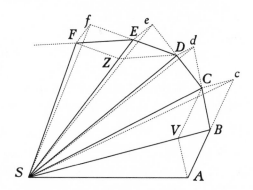

cC be drawn parallel to BS, meeting BC at C; and, the second part of the time being completed, the body (by Corollary 1 of the Laws) will be located at C, in the same plane as the triangle ASB.

Connect SC, and, because of the parallels SB, Cc, triangle SBC will be equal to triangle SBc, and therefore also to triangle SAB. By a similar argument if the centripetal force should act successively at C, D, E, and so on, making the body describe the individual straight lines CD, DE, EF, and so on, in the individual particles of time, all these will lie in the same plane, and the triangle SCD will be equal to the triangle SBC, and SDE to SCD, and SEF to SDE. Therefore, in equal times equal areas are described in a motionless plane; and, *componendo*, any sums whatever of areas $SADS$, $SAFS$ are to one another as are the times of description. Now let the number of the triangles be increased and their breadth decreased *in infinitum*, and their ultimate perimeter ADF (by Corollary Four of Lemma Three) will be a curved line: and therefore the centripetal force, by which the body is perpetually drawn back from the tangent of this curve, will act without ceasing, while any described areas whatever $SADS$, $SAFS$, always proportional to the times of description, will be proportional to those same times in this case.

Q.E.D.

Notes on Book One Proposition 1

One stands in awe of the amazing power of this simple proposition, on which rests all of Newton's celestial mechanics, and indirectly his proofs for universal gravitation. Consider the following aspects of the range of this proposition's applicability.

● **Any Force Law:** This proposition will apply with any force law. Beware of assuming now or later that this somehow depends on an inverse square force law. The forces could be completely independent of the distance of the body (as well as of any other property of the body). Or, if they depend on distance, they could be directly as the distance to some power. They could also vary with time since the creation of the universe or depend on the movement of sunspots or the population of tent caterpillars.

● **Changing Force:** The forces may be different from moment to moment not only in magnitude but also among positive, negative, and zero values. As long as the times are held equal, the triangles will be equal regardless of the values of the forces. Nothing is specified about the forces except that they are centripetal and directed to a center that is immobile. As shown in the proof, the forces can be positive, zero, or negative; they can be constant or varying.

● **Body at the Center:** Nothing is said about any body at the center of forces, and no such body is assumed. The center of forces is a mathematical entity, a geometrical point around which the equal areas may be found. It will be a key discovery in the development of Newton's theory of universal gravitation that in certain circumstances there *are* bodies at this geometrical center of forces.

We must be cautious here, as we always must as we work through this development, not to be assuming what we "know" as a consequence of this book's conclusions having become part of our current world view. Remember that as we go into *Principia* we understand gravity only as terrestrial heaviness. What makes the heavenly bodies orbit has been understood to be another matter entirely.

● **Center of Force, Force at the Center:** We might be inclined to think of the center as a center of *force* (singular) because we are imagining something there exerting that force. But Newton is presenting it as the point towards which all the forces around the path, the forces at each of the infinite numbers of points on the path, are directed, and is consistent in this enunciation, and in all the corollaries, in calling it "center of forces." This is not a consequence of his speaking of *bodies* (plural) in the enunciation, since in Corollaries 1–3 the body is singular.

We might even catch ourselves talking about "the force at the center." But Newton has not given us a force at the center, only a center towards which the forces are directed. On the other hand, during the course of the sketch, he does allow himself to speak about "the centripetal force" (singular) acting with an impulse, acting successively, and (at the limit) perpetually drawing back the body from the tangent to the curve.

The important thing is that none of Newton's proofs in any way assumes or depends upon the force being singular or being exerted from the center. This generality of the proofs has an important significance in the larger system of *Principia*. Newton says he does not know the cause of gravity, and that he contrives no hypotheses about it (see the General Scholium, page 96). Thus we don't know whether the force that turns the planets out of their tangential inertial path is a single force somehow pulling from the center, or many forces pushing from behind the planets, or something different yet. Because we don't know, we need our demonstrations to be general enough to allow for the different possibilities.

• **Orbit:** Although the word "orbit" may suggest a closed path to you, there is no assumption in the proof of this proposition that the path be closed.

• **No "ghost curve":** The bases of triangles in the proposition—the distances the body travels between impulses of force—do not circumscribe, and are not inscribed in, this ultimate curve, nor do they connect to it or follow it in any other way as a kind of "ghost curve." They are more like tangents to the curve-to-be than like chords; but they are not tangents, either, because the curve does not exist as long as the force is impulsive: it exists only at the limit. The points A, B, C,... of the finite case polygon do not necessarily fall anywhere on the ultimate curve, nor do any of the points on the sides such as AB, BC, CD.

• **Accelerative force.** Here and in general throughout Books I and II, when Newton says "force" he means accelerative quantity of force (Definition 7). It is a measure of force proportional to the velocity that is generated by that force in a given time. It is only in Book III that he will begin to work with motive quantity of force (Definition 8), which is measured by the motion generated in a given time, and thus involves mass.

Newton has a complex and difficult structure to build. He begins with the simpler case, one in which mass is ignored. Only when he has established what he can about accelerative force does he take the next step, which is bringing in the mass of the attracted body. This gives us another level of complication. Then, finally, he brings in the third level of complication, the mass of the attracting body. The reader of *Principia* must be grateful to him for building gradually in this way, since, had he not, what is already difficult would have been much more so. These second and third levels of complication are introduced in Book III.

Expansion of Newton's Sketch of I.1

"The areas which bodies driven in orbits [*gyros*] describe by radii drawn to an immobile center of forces, are contained in immobile planes and are proportional to the times."

Given:

Immobile center of forces.

To Prove:

1. The areas that bodies driven in orbits describe by radii drawn to that center of forces are proportional to the times;

2. the path of the body and the center of forces remain in the same plane.

Proof:

Part 1

Step 1: Equal Areas in Equal Times

"Let the time be divided into equal parts, and in the first part of the time let the body, by its inherent force, describe the straight line AB. In the second part of the time, the same body, if nothing were to impede it, would pass on by means of a straight line to c (by Law 1), describing the line Bc equal to AB, with the result that, radii AS, BS, cS being drawn to the center, the areas ASB, BSc would come out equal."

The First Law of Motion stated:

> Every body continues in its state of resting or of moving uni-formly in a straight line, except insofar as it is driven by impressed forces to alter its state.

Suppose that in the first part of the time the body moves from *A* to *B* by its inherent force. That is, it has some velocity at *A* and continues with that same velocity in a straight line unless some added force changes it, according to the First Law of Motion. From *A* to *B* we are assuming no external force is operating on the body.

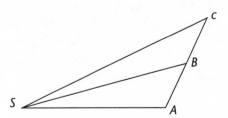

In the second part of the time it would, again by Law 1, if not hindered, move directly in a straight line to *c*, where $Bc = AB$ (since the times are equal and no force has been impressed to change the velocity).

By Euclid I.38 (same height, equal bases), $\triangle SAB = \triangle SBc$. Therefore in this case, the situation or instance of a zero force, equal areas will be described by the radii in equal times.

"But when the body comes to B, let the centripetal force act with an impulse that is single but great, and let it have the effect of making the body depart from the straight line Bc and continue in the straight line BC. Let cC be drawn parallel to BS, meeting BC at C; and, the second part of the time being completed, the body (by Corollary 1 of the Laws) will be located at C, in the same plane as the triangle ASB."

Now suppose a centripetal force acts on the body at B, turning it aside into line Bx. This force is to be understood as a single impulse of significant magnitude operating at that moment. That magnitude is measurable in the amount by which Bx is deflected from its default line Bc. To find that magnitude, draw cy from c parallel to SB, meeting Bx at C.

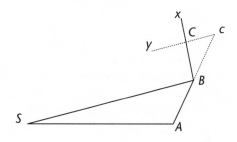

The centripetal force is impelling the body from B towards S. BS is its direction of force. Its magnitude is found by the actual deflection cC that brings the body, in a line parallel to BS, to the actual point C. C is reached in the same time that would have taken the body to c.

The times here for all these motions AB, Bc, BC, and so on, are equal; but the distances traveled AB, BC, and so on, are not necessarily equal.

We will call cC the effect of the force exerted as an impulse on the body at B for purposes of resolving the two forces.

BC is the resolution of the first force Bc (where its innate force is impelling it) and the second force cC (what centripetal force has accomplished in moving it to C instead of c). BC is the body's actual motion in the given time.

Note that although Bc is necessarily equal to AB, BC is not necessarily the same length as Bc. One must find C using the parallel to line BS.

"Connect SC, and, because of the parallels SB, Cc, triangle SBC will be equal to triangle SBc, and therefore also to triangle SAB."

$\triangle SBc = \triangle SBC$ because they are on the same base SB and between the same parallels SB and Cc [Euclid I.37]. Therefore it will

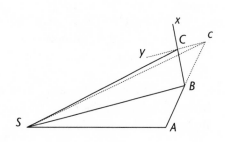

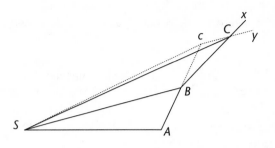

also be true that when we have a positive centripetal force, a nonzero force directed towards the center, equal areas will be described by the radii in equal times.

Suppose finally that we have a repulsive force directed away from the center of forces, a negative centripetal force. *Bx* will now lie on the other side of *Bc*, as will our new point C. However, triangles *SBC* and *SBc* will still lie on the same base between the same parallels and so the areas will be equal. Thus if we had a negative centripetal force, a nonzero force away from the center, we would also have equal areas described in equal times.

We will show in a moment that not only does C lie in the plane of △*SAB* but also the center of forces and all the triangles lie in one plane, as asserted in the enunciation.

"By a similar argument if the centripetal force should act successively at C, D, E, and so on, making the body describe the individual straight lines CD, DE, EF, and so on, in the individual particles of time, all these will lie in the same plane, and the triangle SCD will be equal to the triangle SBC, and SDE to SCD, and SEF to SDE. Therefore, in equal times equal areas are described in a motionless plane;..."

Now suppose this process to be repeated with always the same increments of time. In the equal-time increment (the "individual particle of time") the body will travel the resultant between the motion due to its inherent force and the motion it would have had under the influence of the centripetal force, in a way exactly analogous to our demonstration above. Thus the area of every triangle will remain equal, regardless of how the forces may vary.

Step 2: The Center of Forces and All the Triangles Lie in One Plane

Triangle *SAB* defines a plane, so those three points are in one plane [Euclid XI.2].

A body travelling in a straight line *ABc* will continue in the same plane, so line *Bc* is also in the "immobile plane" [Euclid XI.1].

Since c and S are both in the plane, line cS will be, and so will triangle *SBc*.

Since B is impelled toward S by the centripetal force, the motion from that force cC will be in the plane of B and S, namely the given plane. So C is in that plane, and triangle *SBC* also.

And so on for all successive changes of motion.

Step 3: Proportionality

"...and, *componendo*, any sums whatever of areas $SADS$, $SAFS$ are to one another as are the times of description."

Having demonstrated the equal areas for equal times, we can add together any number of consecutive triangles (having proved in Step 2 that they all lie in one plane) to form polygonal areas; these areas will be proportional to the times of their description.

For example, having shown that

$$\triangle BCS : \triangle ABS :: t_{BC} : t_{AB} \ , \quad \triangle CDS : \triangle ABS :: t_{CD} : t_{AB} \ , \text{etc.,}$$

it follows *componendo* that

$$\triangle ABS + \triangle BCS + \triangle CDS + \cdots \ : \ \triangle ABS \ :: \ t_{ABCDK} : t_{AB} \quad ;$$

that is, the polygonal area *SABCD...* has to the original area *SAB* the same ratio that the time required to move along path *ABCD...* has to the time to traverse *AB*. And so on for any combination of triangles.

Part 2

"Now let the number of the triangles be increased and their breadth decreased *in infinitum*, and their ultimate perimeter ADF (by Corollary Four of Lemma Three) will be a curved line: and therefore the centripetal force, by which the body is perpetually drawn back from the tangent of this curve, will act without ceasing, while any described areas whatever $SADS$, $SAFS$, always proportional to the times of description, will be proportional to those same times in this case.

Q.E.D."

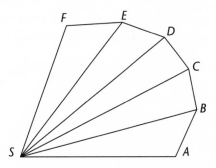

Since any described area (sum of triangles) is proportional to the time, this will also hold for evanescent triangles, the limiting case as the equal times approach zero. This limit of a sum of triangles, as Newton shows in his Lemma 3 Corollary 4, will be a curvilinear area. The triangles themselves, no matter how many there are, will always be a rectilinear area, of course. But the more there are of them the closer they will come, when added together, to filling a curvilinear space.

In this limiting case as well, the areas will be proportional to the times.

Q.E.D.

Note that as the time approaches zero, the forces, while in theory remaining impulsive, approach closer and closer to acting at every point. The limiting case, corresponding to an actual curvilinear path, may be thought of as a continuous force, although that might misleadingly suggest a single force. Perhaps we should say "forces acting continuously turning the body out of its tangential motion."

Pause After Proposition 1

● When we first encountered Definition 3, the definition of inherent force, we asked ourselves in what sense this was a force (or perhaps the question was what force meant to Newton if he included inertia as a force). One might be struck, therefore, to see him using both inherent force and impressed force to calculate the path of the body in this proposition; this seems to be an application in which their homologous effects would justify seeing them as the same sort of entity.

But do note that Newton is not actually using the two sorts of force to yield a resultant force; rather, he is letting each individual sort of force result in a motion and resolving the two motions into a resultant.

● In the *New Astronomy* (1609), Johannes Kepler invented the areas-proportional-to-times rule as a way of mathematizing the observed fact that planets go faster as they get closer to the sun. This later came to be known as Kepler's Second Law. Newton is here deriving the same conclusion starting from a hypothesis of central forces (forces directed radially towards or away from a fixed center). Its agreement with what had been recognized as the real-world applicability of Kepler's rule does not, however, consititute proof that Proposition 1 applies to our world. It is still only hypothetical, and will be applied to our world in Book III. At this point in the unfolding of *Principia,* we don't know whether there are such things as central forces in our world. What this proposition tells us is that if there are such forces operating, the result will be equable description of areas.

Book I Proposition 4:
A Key Step towards Universal Gravitation

If you have gotten this far, you have already experienced some of the power and simplicity of the *Principia*. The next proposition and its corollaries requires a little more work, but but the extra trouble will be very rewarding. It provides the crucial link between Kepler's radius/period relation in planetary orbits and Newton's universal physics. Kepler had shown that in the solar system the periodic times of the planets are proportional to the 3/2 power of their mean distances from the sun. In Book I Proposition 4 and its corollaries, Newton shows that this Keplerian law implies a force that varies inversely as the squares of the distances. Once you have gone through this proof, you will be in a position to establish the universal inverse-square relation for both falling bodies and orbiting planets, which is one of the most important, and certainly the most thrilling, step in Newton's proof of universal gravitation. This takes place in Book III Propositon 4 (included in this module).

Proposition I.4 itself, together with Corollary 6, will be used to show the inverse-square relationship for the moons of Jupiter and Saturn and for the solar system as a whole. Corollary 9 provides a means of comparing the motion of a falling body with the motion of an orbiting body. It will be used in Proposition 4 of Book III to prove that our moon is held in its orbit by the power of gravity, the same familiar yet mysterious power that makes things on earth fall. Having proved that one celestial body is moved by gravity, Newton is in a position to extend the proof to all bodies in the universe. That proof is beyond the scope of this module (it is presented in detail in *Newton's Principia: The Central Argument,* published by Green Lion Press). Nevertheless, the selection included here gives one a taste of how universal gravitation was established.

Book I Proposition 4

The centripetal forces of bodies that describe different circles with uniform motion tend towards the centers of the same circles, and are to one another as the squares of arcs described in the same times applied to [i.e., divided by] the radii of the circles.

[Newton's proof is more complicated than is required, and depends on corollaries to Prop. 1 that we have not proved. A simplified proof, which closely follows Newton's line of reasoning, is substituted below.]

Notes on 1.4

• In this proposition and its first eight corollaries, Newton is moving from given orbits to a corresponding force law governing the bodies in those given orbits. The force law he finds may or may not govern the movements of any other bodies at other distances. He is not, in short, assuming or deducing a field.

• Consider also that nothing in this proposition requires that the orbits have the same centers. One might be comparing the orbit of one of Jupiter's moons and Venus's orbit around the sun. This would indeed give us a ratio of forces, but not anything generalizable to other orbits.

• When Newton speaks of a force (or anything else) varying as one thing directly and something else inversely, we are to understand a simultaneous dependence. He does not mean to make two independent statements. To get the ratio of forces in this example, we must compound the ratio of the first with the inverse ratio of the second. See pages xvii–xviii of Preliminaries for an explanation of compounding ratios.

• Keep in mind that, like all the proofs in Book I of *Principia*, this proposition is hypothetical. That means that, as long as the conditions are met, the proof applies, not just to our universe, but to *any possible universe*. You might want to think about how we could know, just by the power of our own thinking, what rules would apply to something that has never existed, and maybe never will.

Proof of I.4

"The centripetal forces of bodies that describe different circles with uniform motion tend towards the centers of the same circles, and are to one another as the squares of arcs described in the same times applied to [i.e., divided by] the radii of the circles."

Given:

1. Bodies moving around the circumferences of circles;

2. the bodies move with equable motion.

To Prove:

1. Centripetal forces on the bodies will be directed to the centers of the circles.

2. The centripetal forces on bodies in any two circles will be to each other as squares of equal-time arcs in the two different circles divided by the circles' respective radii, even for finite arcs.

That is, the ratio of forces will be as the ratio of the squares of the arcs compounded with the inverse of the radii:

$f_1 : f_2 :: (\text{arc}_1{}^2 : \text{arc}_2{}^2) \text{ comp } (\text{radius}_2 : \text{radius}_1).$

Or, $f \propto \text{arc}^2 / \text{radius}.$

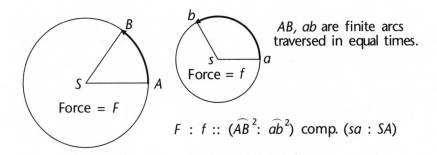

AB, ab are finite arcs traversed in equal times.

$F : f :: (\widehat{AB}^2 : \widehat{ab}^2) \text{ comp. } (sa : SA)$

Proof:

Part 1: Force Directed to Centers of Circles

The bodies are given as moving uniformly around circles. Within each circle we will have equal arcs in equal times because of the uniform motion. The radii within one circle will be equal by the nature of circles. Therefore the sectors will be equal around the center of the circle. Thus, for each circle, the areas described in equal times will be equal and the bodies will describe areas proportional to the times around the centers of the respective circles.

By Proposition 2 (the converse of Proposition 1), if a body describes areas proportional to the times about a point, it is urged by a centripetal force directed to that point.

Part 2: Force ∝ Arc² / Radius

The proposition and sketch state this relationship generally, not just for evanescent (vanishingly small) arcs. Newton first demonstrates it for the infinitesimal case, that is, as the arcs become extremely small, and then he extends it to the finite case.

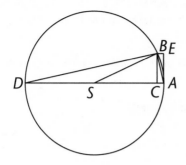
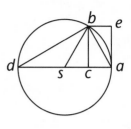

Suppose we have two bodies travelling counterclockwise along circular arcs $\overset{\frown}{AB}$ and $\overset{\frown}{ab}$ around centers S and s in equal times. Draw the diameters AD and ad, and the chords AB, ab, BD, and bd. Draw BC and bc perpendicular to AD and ad, respectively. Draw tangents AE, ae from A and a perpendicular to AD and ad, respectively (Euclid III.16 and Porism), and draw BE, be parallel to DA and da, respectively.

Let's consider what the forces are doing to the bodies (we'll just look at the small-letter version; the same will apply to the capital letter figure).

If no force were acting on the body, it would go from a along the tangent in a straight line to e. The force, however, acting always from a in the general direction of as, pulls the body to the left so that it ends up at b. The difference, line eb, therefore is the effect produced by the force. Since the deflections EB, eb are generated in equal times, their lengths are proportional to the average velocities produced by the forces, which we can treat as constant because the time intervals are very short. (In fact, for circular motion around the center of a circle, the force does remain constant, as this proposition proves). Therefore, by Definition 7, the accelerative forces are proportional to the deflections:

$$F_A : F_a :: EB : eb.$$

Now you will probably object that the deflection eb isn't towards s, so it doesn't really fit the requirements of Law 2. However, as in Proposition 1, if the time interval for the motion ab is made vanishingly small, the line eb (extended as required) comes to be as close as you please to passing through s. So for suitably small arcs, we can take EB and eb as proportional to the forces acting on the two bodies over the arcs AB and ab:

$$F : f :: EB : eb.$$

The little lines AC and ac are clearly the same length as EB and eb, since they form the opposite sides of rectangles.

Therefore,

$$F : f :: AC : ac, \quad \text{or} \quad F \propto AC. \tag{1}$$

Now we need to relate these lines AC, ac to things we can measure.

Consider the three triangles abc, abd, and bcd, and their counterparts in the other diagram.

Euclid proved (III.31) that any triangle that is inscribed in a semicircle (such as ABD and abd here) has a right angle on the circumference (that is, at B and b). And since abc and bcd are also right triangles, and since the acute angles of any right triangle add up to 90 degrees (Euclid I.32), these three triangles have the same three angles, and are all the same shape: they are *similar* triangles.

When geometrical figures are similar, their sides are proportional, so that if (for example) ab were twice, three times, etc., as long as ac, then ad (in triangle abd) would also be twice, three times, etc., as long as ab. We can express this in fractions:

$$\frac{ac}{ab} = \frac{ab}{ad} \quad \text{and} \quad \frac{AC}{AB} = \frac{AB}{AD} .$$

Multiply both sides of these equations by ab and AB, respectively; we have

$$ac = \frac{ab^2}{ad} \quad \text{and} \quad AC = \frac{AB^2}{AD} .$$

Substituting this into Equation 1,

$$F : f :: \frac{AB^2}{AD} : \frac{ab^2}{ad} , \quad \text{or} \quad F \propto \frac{AB^2}{AD} .$$

Remember, this is true only when the time of travel is very short, and arcs AB and ab become vanishingly small. But in that case, the arc gets as close as you please to being the same length as its chord (Newton proves this in Lemma 7; we shall take it as a plausible assumption). Therefore, we may substitute the arcs for the chords:

$$F : f :: \frac{\widehat{AB}^2}{AD} : \frac{\widehat{ab}^2}{ad} , \quad \text{or} \quad F \propto \frac{\widehat{AB}^2}{AD} .$$

Now AD is twice AS, and ad twice as, so

$$F : f :: \frac{\widehat{AB}^2}{2AS} : \frac{\widehat{ab}^2}{2as} :: \frac{\widehat{AB}^2}{AS} : \frac{\widehat{ab}^2}{as} .$$

We are given that the velocity on each circle is constant. Let V be the velocity on circle ABD, and v be the velocity on abd. Then

arc $AB = V \times t$, and arc $ab = v \times t$,

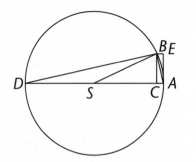 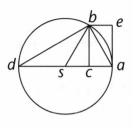

where t is the (very short) time to traverse the arcs. Therefore,

$$\widehat{AB}:\widehat{ab} :: V{\times}t:v{\times}t :: V:v \tag{2}$$

This is a fixed ratio, valid for all values of t. So we no longer have to keep the arcs very small. Therefore, as Newton says,

The centripetal forces . . . are to one another as the squares of arcs described in the same times applied to [i.e., divided by] the radii of the circles.

That is,

$$F:f :: \frac{\widehat{AB}^2}{AS}:\frac{\widehat{ab}^2}{as} \quad \text{or} \quad F:f :: \frac{\widehat{AB}^2}{R}:\frac{\widehat{ab}^2}{r}$$

where we have written R and r in place of radii AS and as, respectively.

Therefore, the forces are proportional to the squares of the arcs divided by the radii.

In **Corollary 1** to this proposition, Newton restates his conclusion in terms of velocities rather than arcs. By equation (2), the velocities are proportional to the arcs, and so we may substitute the squares of the velocities for the squares of the arcs. Therefore,

$$F:f :: \frac{V^2}{R}:\frac{v^2}{r} \ . \tag{3}$$

In **Corollary 2**, Newton again restates the relation, this time in terms of periodic times. Since the time to complete one orbit around the circle is the length of the circumference (2π times the radius) divided by the velocity, in the circle of radius R we have

$$P^2 = \frac{4\pi^2 R^2}{V^2}, \quad \text{therefore} \quad V^2 = 4\pi^2 R^2 / P^2$$

where P is the periodic time. Similarly for the circle of radius r,

$$v^2 = 4\pi^2 r^2 / p^2$$

where p is the periodic time. Substituting these expressions for V^2 and v^2 into proportion (3),

$$F:f \;::\; \frac{4\pi^2 R^2/P^2}{R} : \frac{4\pi^2 r^2/p^2}{r} \;::\; \frac{4\pi^2 R}{P^2} : \frac{4\pi^2 r}{p^2} \;::\; \frac{R}{P^2} : \frac{r}{p^2} \;,$$

that is,

$$F \propto \frac{R}{P^2}. \qquad\qquad (4)$$

In **Corollaries 3–7**, Newton finds the ratios of forces for a variety of radii and periodic times. Corollary 6 is especially important, since it will be applied to our universe. It states (in terms that have been modernized),

If the periodic times be in the ratio of the 3/2 power of the radii, and therefore the velocities in the ratio of the square roots of the radii, the centripetal forces will be inversely as the squares of the radii, and conversely.

This is easily obtained by substituting $R^{3/2}$ for P in Equation (4).

I.4 Corollary 9

From the same demonstration it also follows that the arc, which a body describes by revolving uniformly in a circle by a given centripetal force in any time whatever, is the mean proportional between the diameter of the circle and the descent traversed by the body falling under the same given force and in the same time.

Expansion of Newton's Sketch of I.4 Corollary 9

Given:

1. Circle *AFD* with diameter *AD* and center *S*, with the center of forces at the center of the circle.

2. A body describes arc *AF* in given time t in equable motion around center of forces *S*.

3. The same body falls from the same initial point *A* towards the same center of forces *S* covering distance *AL* in the same time t.

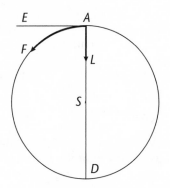

To Prove:

The arc traveled by the body in circular motion is a mean proportional between the circle's diameter and the distance fallen in the same time.

That is, $AD : \overset{\frown}{AF} :: \overset{\frown}{AF} : AL.$

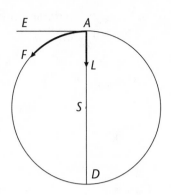

Proof:

Step 1: The Ratio in the Finite Case

The distance traveled around the circle in equable motion is a function of time. According to Galileo, in *Two New Sciences*, Proposition 4 of Equable Motion, distances traveled in equable motion vary as the speeds times the times.

arc $\propto v \times t$.

So since v is constant, arc $\propto t$, and

$$\text{arc}^2 \propto t^2. \tag{1}$$

By Galileo, *Two New Sciences*, Proposition 2 of Naturally Accelerated Motion,

> If a moveable descends from rest in uniformly accelerated motion, the spaces run through in any times whatever are to each other as the duplicate ratio of their times; that is, are as the squares of those times.

Thus,

distance fallen $\propto t^2$. $\qquad\qquad$ (2)

Combining Equations 1 and 2,

$$\text{arc}^2 \propto \text{distance fallen.} \tag{3}$$

This proportionality will hold for any two arcs and corresponding distances fallen, even if one is evanescent.

Let's see what the relationship between those two magnitudes is at the limit as the given time approaches zero. We will then know the relationship in all cases.

Step 2: Finding the Relationship in the Ultimate Case

The conclusion of Step 2 must be understood as applying only to the limiting case at the very beginning of motion.

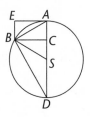

In the course of proving the main part of the proposition, we proved that

$$AC : AB :: AB : \text{diameter } AD.$$

We also used Newton's statement (proved elsewhere in *Principia*) that "ultimately," that is, for evanescent or vanishingly small arcs, the arc is equal to the chord. Therefore

$$AC : \overset{\frown}{AB} \overset{ult}{::} \overset{\frown}{AB} : \text{diameter } AD. \tag{4}$$

Step 3: Conclusion

Now we put what we have discovered about the ultimate case together with what we could say generally.

From equation (3) above, the square of the arc is proportional to the distance fallen, for both large and tiny arcs. In the adjacent diagram,

$$\frac{AL}{AC} = \frac{\overset{\frown}{AF}^2}{\overset{\frown}{AB}^2}$$

and from equation (4),

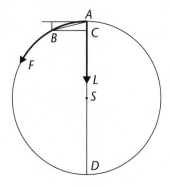

$$\frac{AC}{\overset{\frown}{AB}} \overset{ult}{=} \frac{\overset{\frown}{AB}}{AD}.$$

Multiplying these two,

$$\frac{AL}{AC} \times \frac{AC}{\overset{\frown}{AB}} \overset{ult}{=} \frac{\overset{\frown}{AF}^2}{\overset{\frown}{AB}^2} \times \frac{\overset{\frown}{AB}}{AD}$$

or

$$\frac{AL}{\overset{\frown}{AB}} \overset{ult}{=} \frac{\overset{\frown}{AF}^2}{\overset{\frown}{AB} \times AD}.$$

Multiplying both sides by $\widehat{AB}/\widehat{AF}$,

$$\frac{AL}{\widehat{AF}} \overset{ult}{=} \frac{\widehat{AF}}{AD}.$$

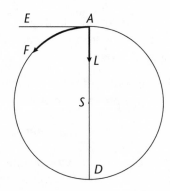

Notice that all of these quantities are finite: the relationship holds true for all times and arc lengths. Therefore, we can drop the "ult."

$$\frac{AL}{\widehat{AF}} = \frac{\widehat{AF}}{AD}.$$

Restated,

$AL: \widehat{AF} :: \widehat{AF} : AD.$

Therefore the arc, which a body describes by revolving uniformly in a circle by a given centripetal force in any time whatever, is the mean proportional between the diameter of the circle and the descent traversed by the body falling under the same given force and in the same time.

Q.E.D.

Book III

On the Syſtem of the World. ❦ Third Book.

Newton's Preface to Book III

In the preceding books I have presented the principles of philosophy, not, how-ever, philosophical, but only mathematical; that is, those from which one can argue in philosophical matters. These are the laws and conditions of motions and forces, which pertain to philosophy in the highest degree. Nevertheless, so that these should not appear sterile, I have illustrated them with certain philosophical scholia, treating those things that are most general, and in which philosophy seems most to be poured out, such as the density and resistance of bodies, spaces void of bodies, and the motion of light and of sounds. What remains is that we show from the same principles the constitution of the system of the world. Concerning this theme I had composed the third book using a popular method, so that it might be read by many. But those by whom the established principles might not have been sufficiently well understood will have a very slight percep-tion of the force of the argument, and will not cast aside prejudices to which they have been accustomed for many years. Therefore, so that the matter not be subject to dispute, I have carried over the substance of that book into proposi-tions, in the mathematical manner, so that they may only be read by those who had previously gone through the principles. Nevertheless, because many propositions appear there [i.e., in the first two books] which might take too much time even for those learned in mathematics, I do not wish to propose that everyone read all of them: it would be sufficient for one to read carefully the definitions, laws of motion, and the first three sections of the first book, and then to carry on to this book on the system of the world, looking up at will the remaining propositions of the previous books cited here.

Rules of Philosophizing

Notes on Rules of Philosophizing

- Newton, writing this book in Latin, called these "Regulae Philosophandi."
In English, rules of philosophizing, or rules of doing philosophy.

- What are these "rules" intended to be?
For a start, we note that Newton has *not* said that these are rules governing the study of some subject matter. Second, we note that, although the title of this book is *Mathematical Principles of Natural Philosophy,* he has not called this section rules for doing *natural* philosophy but rules for doing philosophy. Perhaps he means to offer this as a set of principles with more general application; perhaps he would apply them to any branch of philosophical thinking.

Rather than intending these "rules" to be something imposed from outside to direct study of a body of knowledge, it seems that Newton sees them as principles that describe the way we actually think if we are thinking philosophically. Applied to natural philosophy, they are standards of sound reasoning about phenomena, causes, and properties of matter. They describe the working of the mind of a careful thinker—in this application, the working of the mind of a natural philosopher, what we would call scientific thinking.

Rule 1

That there ought not to be admitted any more causes of natural things than those which are both true and sufficient to explain their phenomena.

Philosophers state categorically: Nature does nothing in vain, and vain is that which is accomplished with more that can be done with less. For Nature is simple, and does not indulge herself in superfluous causes.

Rule 2

Accordingly, to natural effects of the same kind the same causes should be assigned, as far as possible.

As, for example, respiration in humans and in animals, the descent of stones in Europe and in America, light in a cooking fire and in the sun, the reflection of light on earth and in the planets.

Rule 3

The qualities of bodies that do not suffer intensification and remission, and that pertain to all bodies upon which experiments can be carried out, are to be taken as qualities of bodies universally.

For the qualities of bodies are apprehended only through experience, and are accordingly to be declared general whenever they generally square with experiments; and those which cannot be diminished cannot be removed. It is certain that against the tenor of experiments, dreams are not to be rashly contrived, nor is a retreat to be made from the analogy of nature, since she is wont to be simple and ever consonant with herself. The extension of bodies is apprehended only through the senses, nor is it perceived in all things. But because it belongs to all perceptible bodies, it is affirmed to be universal. We experience many bodies to be hard. Hardness of the whole, moreover, arises from hardness of the parts, and thence we rightly conclude that the undivided particles, not only of these bodies which are perceived, but also of all others, are hard. We conclude that all bodies are impenetrable, not by reason, but by perception. Those that we handle are found to be impenetrable, and thence we conclude that impenetrability is a property of bodies universally. That all bodies are moveable, and that by certain forces (which we call the forces of inertia) they persevere in motion or rest, we gather from these very properties in bodies that we see. Extension, hardness, impenetrability, moveability, and the force of inertia of the whole arise from the extension, hardness, impenetrability, moveability, and forces of inertia of the parts, and thence we conclude that all the least parts of all bodies are extended and hard and impenetrable and moveable and endowed with forces of inertia. And this is the foundation of all of philosophy. Moreover, that parts of bodies that are divided and mutually contiguous can be separated from each other, we come to know from the phenomena, and that undivided parts can by reason be divided up into smaller parts, is certain from mathematics. But whether those distinct and hitherto undivided parts can be divided by the forces of nature and separated from each other, is uncertain. And if it were to be established by but a single experiment that by breaking a hard, solid body, some undivided particle were to suffer division, we would conclude, by the force of this rule, not only that the divided parts are separable, but also that the undivided ones can be divided *in infinitum*.

Finally, if it be established universally by experiments and astronomical observations that all bodies on the surface of the earth are heavy towards the earth, and this according to the quantity of matter in each, and that the moon is heavy towards the earth according to the quantity of its matter, and that our sea in turn is heavy towards the moon, and that all the planets are heavy towards each other, and that the gravity of comets towards the sun is similar, it will have to be said, by this rule, that all bodies gravitate towards each other. For the argument from phenomena for universal gravitation will be even stronger than that for the impenetrability of bodies, concerning which we have absolutely no experiment in the heavenly bodies; nay, not even an observation. Nevertheless, I do not at all assert that gravity is essential to bodies. By "inherent force" [*vis insita*] I understand only the force of inertia. This is inalterable. Gravity is diminished in receding from earth.

Rule 4

In experimental philosophy, propositions gathered from the phenomena by induction are to be taken as true, whether exactly or approximately, contrary hypotheses notwithstanding, until other phenomena appear through which they are either rendered more accurate or liable to exceptions.

This must be done lest an argument from induction be nullified by hypotheses.

Phenomena

Notes on Phenomena

- The following six "Phenomena" are the transition between the hypothetical propositons of Books I and II and the propositions of Book III that
make application to our world and draw conclusions about how things work
in our world.

- **Calculations and Planetary Theory.** Despite what might be suggested
by their title, most of these "Phenomena" invoke not just observations, but
planetary theory in current use by the astronomers of his time. Newton
doesn't spell all that out; he assumes we are knowledgeable about contemporary planetary theory. Because this can be complex, I have omitted the
demonstrations of all but one of the Phenomena.

- Phenomenon 3, which demonstrates that the planets move around the
sun, thus refuting a strictly earth-centered hypothesis, requires no planetary
theory, only some careful thinking and visualization. It is therefore included,
along with some aids in that visualization.

Phenomenon 1

---

*That the planets around Jupiter, by radii drawn to the center of Jupiter,
describe areas proportional to the times, and that their periodic times
(the fixed stars being at rest) are in the sesquiplicate ratio of the
distances from Jupiter's center.*

This is established from astronomical observations. The orbits of these planets do not
differ perceptibly from circles concentric about Jupiter, and their motions in these
circles are found to be uniform. But that the periodic times are in the sesquiplicate ratio
of the semidiameters of the orbits, the astronomers are in agreement. . .

[The demonstration is omitted.]

Phenomenon 2

That the planets around Saturn, by radii drawn to Saturn, describe areas proportional to the times, and that their periodic times, the fixed stars being at rest, are in the sesquiplicate ratio of the distances from Saturn's center.

[The demonstration is omitted.]

Phenomenon 3

The five primary planets, Mercury, Venus, Mars, Jupiter, and Saturn, enclose the sun in their orbits.

That Mercury and Venus revolve around the sun is demonstrated from their lunar phases. When they shine with a full face, they are located beyond the sun, when halved they are even with the sun, and when sickle-shaped they are this side of the sun, sometimes passing across its disk in the manner of sunspots. Further, from Mars's full face near conjunction with the sun, and its gibbous face in the quadratures, it is certain that it encompasses the sun. The same is also demonstrated of Jupiter and Saturn from their ever full phases, for it is manifest from the shadows of satellites cast upon them that these shine with light borrowed from the sun.

Notes on Phenomenon 3

● The assertion of this Phenomenon does not include a claim that the earth moves around the sun. The sun, with its five encompassing planets, could revolve around the earth. This was the theory of the Danish astronomer Tycho Brahe (1546–1601).

● In order to think through the argument of this Phenomenon, we must understand how a planet would look given different angles between us observing from earth, the sun providing the light, and the planet.

Lemmita: How the Planet Looks

The following assumes that the light we see from the planets is reflected light of the sun. If we believed that the light might be coming from the planet, all this would change. In that case, however, it would be hard to make sense of the phase observations for Mercury, Venus and Mars. And Newton ends this Phenomenon with a demonstration that the light we see from Jupiter and Saturn is borrowed from the sun.

Given: Sun S, earth E, and planet P.

To Prove:

1. If angle *EPS* is right, the planet will show as half phase.

2. If angle *EPS* is acute, the planet will show as gibbous.

3. As angle *EPS* → 0, phase → full.

4. If angle *EPS* is obtuse, the planet will show as crescent.

Proof:

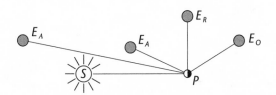

We will consider possibilities for any configuration of sun, earth, and planet.

We may have the earth at points E_A where the angle *EPS* is acute, either with the earth farther from the sun than the planet, or nearer.

Or, we may have the earth at a point such as E_R, at whatever distance, such that angle *EPS* is a right angle.

Finally, we may have the earth at a point such as E_O, at whatever distance, such that angle *EPS* is obtuse.

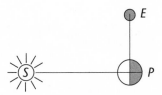

Case 1: First Consider Right Angle *EPS*

What do we see? (We must imagine this in three dimensions, and shift from the diagram's view from above the three bodies to a view from the surface of the earth.)

We would see half phase.

Case 2: Now Consider Acute Angle *EPS*

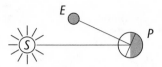

Adding the third dimension and shifting 90° to look at this from our earth's eye view, what do we see?

We would see gibbous.

Case 3: What Happens when Acute Angle is Nearly 0°?

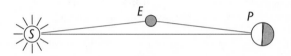

What would we see from the earth?

We would see the planet as nearly full.

Case 4: Now Consider an Obtuse Angle *EPS*

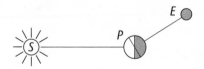

Shifting 90° into the paper to imagine this viewed from the earth, what would we see?

We would see a crescent.

Q.E.D.

Expansion of Newton's Sketch of Phenomenon 3

To Prove:

1. That Mercury and Venus revolve around the sun.

2. That Mars encompasses the sun.

3. That the same is demonstrated for Jupiter and Saturn.

Proof:
Part 1: Mercury and Venus

"That Mercury and Venus revolve around the sun is demonstrated from their lunar phases. When they shine with a full face, they are located beyond the sun, when halved they are even with the sun, and when sickle-shaped they are this side of the sun, sometimes passing across its disk in the manner of sunspots."

We must take as given the following additional observational conclusion Newton doesn't cite, but which was part of the common astronomical knowledge of the time:

General Observation:

The maximum elongation of Venus and Mercury is a small acute angle ($48°$ for Venus, even less for Mercury). The elongation is the angle between the sun and the planet as we view it from the earth, angle SEP in these diagrams.

Thus we know for these two planets that angle SEP is always acute.

We also may consider the following particular observations, which Newton does cite:

Particular Observations:

Observation 1: We sometimes see Mercury and Venus near to full.
Observation 2: Sometimes we see them half full.
Observation 3: At other times they are seen as sickle-shaped (crescent).
Observation 4: Sometimes the planets are seen like spots traversing the sun's disk.

Step 1:

"That Mercury and Venus revolve around the sun is demonstrated from their lunar phases. When they shine with a full face, they are located beyond the sun..."

Now suppose by Observation 1 we find the planet near to full.

What must the configuration be then?

It could be close to full in this configuration:

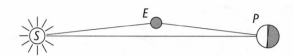

By the lemmita entitled How the Planet Looks, in the notes above, if phase is approaching full, angle *EPS* is approaching 0, as it does in this configuration.

But in this configuration angle *SEP* would not be small acute, contrary to the General Observation above.

Therefore this isn't the configuration.

It would also be near to full in this configuration:

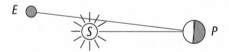

Here angle *EPS* → 0 as required by the lemmita. But here angle *SEP* also → 0.

Therefore, when we see Mercury or Venus near to full, the planet is beyond the sun.

Q.E.D. for Step 1 of Part 1.

Step 2

"...when halved they are even with the sun, ..."

By Observation 2, sometimes they are half full.

What configuration would give us this?

By the lemmita, angle *EPS* = a right angle.

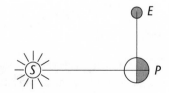

The diagram below illustrates the two possible configurations, which are mirror images of one another. In both cases angle *SEP* is acute, consistent with the General Observation above.

Therefore, from the point of view of the earth, when the planet is half full, it is "even with the sun," to one side or the other, as shown:

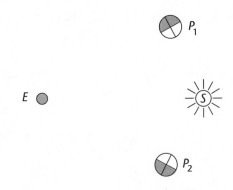

Q.E.D. for Step 2 of Part 1.

Step 3a

"...and when sickle-shaped they are this side of the sun,..."

By the lemmita, the planet would appear sickle-shaped or crescent when angle *EPS* is obtuse.

This can only happen in this configuration:

Here angle *SEP* is still acute, so it fits the General Observation.

Therefore when the planet is seen as crescent, it is on our side of the sun, between the sun and earth.

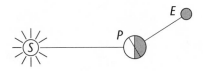

Q.E.D. for Step 3a of Part 1.

Step 3b

"…sometimes passing across its disk in the manner of sunspots."

By Observation 4, the planets sometimes are seen like spots traversing the sun's disk.

In this case again they must be passing between the sun and earth.

Q.E.D. for Step 3b of Part 1.

Conclusion of Part 1

Thus we have seen that Venus and Mercury are sometimes between us and the sun (Steps 3a and 3b), sometimes at the sides of the sun (Step 2) and sometimes beyond the sun (Step 1).

From this we conclude that Mercury and Venus revolve around the sun.

Q.E.D.

Part 2: Mars Encompasses the Sun

"Further, from Mars's full face near conjunction with the sun, and its gibbous face in the quadratures, it is certain that it encompasses the sun."

Step 1

"…full face near conjunction with the sun, …"

We observe that Mars is near full when it is near conjunction, that is, when it is on the same side of the earth as the sun. What configurations would put it on the same side?

Here is one configuration in which Mars and the sun are on the same side of the earth:

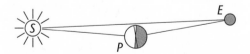

This is conjunction, but it wouldn't be full here, but "new." Angle *EPS* is obtuse, and by the lemmita, the planet would be crescent.

In the configuration shown below, Mars and the sun are also on the same side.

Angle *EPS* is acute approaching 0°; therefore Mars's phase is approaching full.

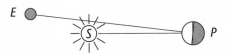

Therefore it must be this configuration; Mars is *beyond* the sun.

Q.E.D. for Step 1 of Part 2.

Step 2

"…and its gibbous face in the quadratures, …"

We observe also that when Mars is at quadrature, that is, when angle *SEP* = 90°, it is gibbous. By the lemmita, angle *EPS* would be acute.

From its quadrature configuration we know that distance *SP* oppo-site the right angle is the greatest side of the triangle and Mars is farther from the sun than the earth is from the sun.

Because Mars comes to quadrature, it is (at least at that point) farther from the sun than the earth is.

Q.E.D. for Step 2 of Part 2.

Conclusion of Part 2

"…it is certain that it encompasses the sun."

Thus we see that Mars is sometimes to the side of us (Step 2) and sometimes on the other side of the sun from us.

Therefore Mars's orbit encompasses the sun.

Q.E.D. for Part 2.

Part 3: Jupiter and Saturn

"The same is also demonstrated of Jupiter and Saturn from their ever full phases, ..."

Our observation here is that they are full in all situations.

Consider conjunction, quadrature, and opposition.

Step 1: Conjunction

At conjunction Jupiter sets with the sun. There are two cases:

In the first case, angle *EPS* is small acute. By the lemmita, Jupiter could look full.

In the second case, angle *EPS* is obtuse. By the lemmita above, Jupiter wouldn't look full but crescent.

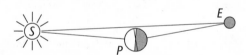

Thus if it looks full at conjunction, it must be in the first configuration.

Therefore Jupiter must be beyond the sun at conjunction.

Q.E.D. for Step 1 of Part 3.

Step 2: Quadrature

At quadrature angle *SEP* = 90°. The observation is that in this configuration Jupiter looks full.

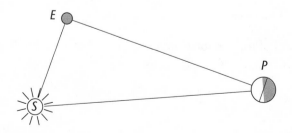

In Part 2, when Mars was at at quadrature, we said that it would appear gibbous. Jupiter, however, is so far away that it appears full even when angle $SEP = 90°$. This is because EP and SP are distances so great in relation to SE that angle EPS is small acute and so by the lemmita Jupiter would appear near to full. It is not visibly gibbous.

Q.E.D. for Step 2 of Part 3.

Step 3: Opposition

At opposition Jupiter rises as the sun sets.

Angle $EPS \rightarrow 0$ and the planet is full if it is on the opposite side of the earth from the sun.

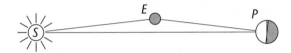

(If Jupiter fell between earth and sun, angle EPS would be obtuse and we would not see the planet full.)

Q.E.D. for Step 3 of Part 3.

Conclusion of Part 3

Therefore, since they are sometimes beyond the sun and sometimes on our side of the sun, Jupiter and Saturn encompass the sun.

Q.E.D. for Part 3.

(See note at the end of Part 1.)

"...for it is manifest from the shadows of satellites cast upon them that these shine with light borrowed from the sun."

We see the moons of Jupiter and Saturn crossing the disks of the planets leaving shadows such that we can see both the moon and the shadow. This might be best seen at quadrature; the moon crosses into Jupiter before the shadow or vice versa.

This proves that Jupiter and Saturn are not their own light source.

Phenomenon 4

That the periodic times of the five primary planets, and that of the sun around the earth or of the earth around the sun (the fixed stars being at rest), are in the sesquiplicate ratio of the mean distances from the sun.

This ratio discovered by Kepler is acknowledged by all. The periodic times are necessarily the same, as well as the dimensions of the orbits, whether the sun revolve around the earth or the earth around the sun. And concerning the measure of the periodic times, there is unanimity among all astronomers.

[The demonstration is omitted. The mean distances and periodic times are easily obtained from a variety of reference sources, and simple computations will show the truth of Newton's claim.]

Phenomenon 5

That the primary planets, by radii drawn to the earth, describe areas by no means proportional to the times, but by radii drawn to the sun, do traverse areas proportional to the times.

For with respect to the earth, they now progress, now are stationary, now even retrogress; but with respect to the sun, they always progress, and do so with a motion very nearly uniform, but a little faster at perihelia and slower at aphelia, so as to make the description of areas equable. The proposition is very well known to astronomers, and in Jupiter particularly it is demonstrated by eclipses of the satellites, by which eclipses we have said that the heliocentric longitudes of this planet and its distances from the sun are determined.

Phenomenon 6

That the moon by a radius drawn to the center of the earth describes an area proportional to the time.

[The demonstration is omitted.]

Book III Propositions

Proposition 1

That the forces by which the planets around Jupiter are perpetually drawn back from rectilinear motions and held back in their orbits, look to the center of Jupiter, and are inversely as the squares of the distances from the same center.

The former part of the proposition is clear from the first Phenomenon and the second or third Proposition of the first Book, and the latter part is clear from the first Phenomenon and the sixth Corollary of the fourth Proposition of the same Book.

The same is understood of the planets that accompany Saturn, from the second Phenomenon.

Expansion of Newton's Sketch of III.1

"That the forces by which the planets around Jupiter are perpetually drawn back from rectilinear motions and held back in their orbits, look to the center of Jupiter, and are inversely as the squares of the distances from the same center. ... The same is understood of the planets that accompany Saturn..."

To Prove:

1. That the forces on the moons tend towards Jupiter's center;

2. that the forces on the moons are inversely as the squares of the distances from Jupiter.

3. Both parts may be shown for Saturn as well.

"The former part of the proposition is clear from the first Phenomenon and the second or third Proposition of the first Book,..."

Proof:
Part 1

By Phenomenon 1, the moons sweep out areas proportional to the times about Jupiter as a center.

If Jupiter were at rest or moving with uniform rectilinear motion we could use Proposition I.2, and would be able to conclude that the center of Jupiter is the center of forces towards which the moons are being impelled by the centripetal force.

Proposition I.2 says:

> Every body that moves in some curved line described in a plane, and, by a radius drawn to a point that either is immobile or proceeds uniformly in a straight line, describes areas about that point proportional to the times, is urged by a centripetal force tending to the same point.

But Jupiter is not at rest, nor is it even moving uniformly in a straight line. Jupiter is, by Phenomenon 3, encompassing the sun in its own curved orbit. (We don't at this point know whether the sun might be at any kind of center for Jupiter's orbit; it's enough to know that to encompass the sun Jupiter will not be moving in a straight line.)

So we must go to Proposition I.3 for our situation here. I.3 says:

> Every body that, by a radius drawn to the center of another body moved in any way whatever, describes areas about that center proportional to the time, is urged by a force compounded of a centripetal force tending toward that other body, and of all the accelerative force by which that other body is urged.

We are told here that the moons, which describe areas around the center of Jupiter proportional to the times, are urged by forces compounded of those which impel them to Jupiter and all the forces by which Jupiter is urged.

If we subtract out the common accelerative force which is impelling a moon-Jupiter system, we are left with the force which is impelling that moon toward the center of Jupiter.

Part 2

"...and the latter part is clear from the first Phenomenon and the sixth Corollary of the fourth Proposition of the same Book."

The "latter part" is Part 2 of our to-prove: that the forces on the moons are inversely as the squares of the distances from Jupiter.

Part 1 of Phenomenon 1 shows that Jupiter's moons travel in equable circular concentric orbits. (We ignore the motion common to the Jupiter-moons system.) This means that we can invoke Proposition I.4, which deals with motion in such circles.

Part 2 of Phenomenon 1 shows that their periodic times are as the $\frac{3}{2}$ powers of their radii. This means that we can use I.4 Corollary 6, which says that, if the periodic times vary as the $\frac{3}{2}$ powers of the radii, the centripetal forces will be inversely as the squares of the radii.

We now know something about the centripetal force causing the moons of Jupiter to depart from their tangential rectilinear inertial motion and curve into orbits around Jupiter: we know that the force acts inversely as the square of the distances of the moons from the center of Jupiter.

That is,

$$f \propto \frac{1}{SP^2} \ .$$

Part 3

"The same is understood of the planets that accompany Saturn, from the second Phenomenon."

We may follow the same reasoning to the same conclusion for the moons of Saturn, substituting the conclusions of Phenomenon 2 for those of Phenomenon 1.

Therefore, for Saturn's moons as well as Jupiter's,

$$f \propto \frac{1}{SP^2} \ .$$

Q.E.D.

Proposition 2

That the forces by which the primary planets are perpetually drawn back from rectilinear motions, and are held back in their orbits, look to the sun, and are inversely as the squares of the distances from its center.

[The demonstration is omitted. It is essentially the same as the demonstration of Proposition III.1, using I.2 and Phenomenon 5 to establish the sun as center of forces and I.4 and Phenomenon 4 to prove the inverse square proportion.]

Proposition 3

That the force by which the moon is held back in its orbit looks to the earth, and is inversely as the square of the distance of places from its center.

[The demonstration is omitted. Because the earth has only one moon, the proof is different from that of the first two propositions. It is, however, too technically difficult to include here.]

III.4, the Famous "Moon Test"

The next proposition this module presents is a crucial proposition in the development of universal gravitation and our modern cosmology, an important milestone in the history of science, and perhaps the most thrilling demonstration in *Principia*.

Remember that, before *Principia*, gravity has meant "terrestrial heaviness" and will continue to mean no more than that until we can show that it does. The enunciation asserts that the moon is drawn off from its tangential motion by force of gravity.

It must be appreciated that this is not the same thing we have been saying all along when we spoke of a centripetal force. It is in fact an assertion that the same power that makes rocks fall on earth is responsible for keeping the moon in orbit.

This notion was startling, and really quite new. And it is still somewhat startling today if one really thinks about it, that is, sets aside what one "knows" because one has been told or read certain things. Look up at that orb hanging in the sky! A rock? It's too round, too luminous! It can be heart-stoppingly beautiful. It has a poetical, even mystical power that has never ceased to have an effect on poets, lovers, and werewolves, despite Newton's work.

Traditionally, the matter of celestial bodies had been regarded as different in kind from terrestrial matter, and this created fundamentally different expectations of the moon and a terrestrial rock. Then, in addition, most theories of celestial mechanics had accounts of why and how the bodies moved in orbits, accounts that relied on very different propensities and mechanisms.

Descartes and Kepler each saw the matter of the moon and the earth as similar and so were in that sense closer to seeing things in the revolutionary way Newton did. But in both cases their ideas of gravity were significantly different from Newton's. Kepler understood terrestrial heaviness as a tendency for like matter to clump together, and he saw the moon (but nothing beyond) as matter like the earth's. However, for Kepler, the moon was retained in its orbit not by that tendency of like matter to clump together, but because that's where God told it to be: it was given its proper distance from the earth according to archetypal principles, and it was given a velocity such that it plays its part in a harmonious whole with other motions in the universe. He would have been far from subscribing to Newton's view.

Descartes saw both the orbit of the moon and the descent of rocks as consequences of the workings of vortices: differential pressures of subtle fluids. This hypothesis encompassed an enormous range of phenomena. It was nevertheless unable to provide quantitative predictions, and mathematicians such as Huygens and Leibniz who tried to get the details of planetary motion out of it failed. (For Newton's remarks on this problem, see the General Scholium at the end of Book III.)

Galileo also believed the moon was a rock, but this was because it *looked* like a rock: for example, light reflected off it the way light reflected off a rock.

Newton, in contrast, gave us, in III.4 here, both the startling claim that the moon is held in orbit by gravity and proof of that claim. It is the first step of the richly-articulated product of rational mechanics that is his theory of universal gravitation as laid out in this book—a theory he uses to predict the regular motions of celestial bodies and all their known anomalies. This is the understanding of gravity that has become a basic component of our common-sense understanding of our physical universe.

Proposition 4

That the moon gravitates towards the earth, and is always drawn back from rectilinear motion, and held back in its orbit, by the force of gravity.

The moon's mean distance from the earth in the syzygies, in terrestrial semidiameters, is 59 according to Ptolemy and most astronomers; 60 according to Wendelin and Huygens, $60\frac{1}{3}$ according to Copernicus, $60\frac{2}{5}$ according to Streete, and $56\frac{1}{2}$ according to Tycho. ...* Let us assume that the mean distance is sixty semidiameters at the syzygies, and that the lunar period with respect to the fixed stars amounts to 27 days, 7 hours, and 43 minutes, as is stated by the astronomers; and that the circumference of the earth is 123,249,600 Paris feet, as is established by the measuring Frenchmen. If the moon be supposed to be deprived of all motion and dropped, so as to descend towards the earth, under the influence of all that force by which (by Proposition 3 Corollary) it is held back in its orbit, it will in falling traverse $15\frac{1}{12}$ Paris feet in the space of one minute. This conclusion comes from a computation based either upon Proposition 36 of the first Book or (what amounts to the same thing) the ninth Corollary of the fourth Proposition of the same Book. For the versed sine of that arc which the moon in its mean motion describes in the time of one minute at a distance of sixty terrestrial semidiameters, is about $15\frac{1}{12}$ Paris feet, or more accurately, 15 feet 1 inch and $1\frac{4}{9}$ lines. Whence, since in approaching the earth that force increases in the inverse of the duplicate ratio of the distance, and is thus greater at the surface of the earth by 60×60 parts than at the moon, a body, in falling by that force in our regions, ought to describe a space of $60\times60\times15\frac{1}{12}$ Paris feet in the space of one minute, and in the space of one second, $15\frac{1}{12}$ feet, or more accurately, 15 feet 1 inch and $1\frac{4}{9}$ lines. And heavy bodies on earth do in fact descend with

---

* Newton here includes an explanation that Tycho's number was based on an error, and shows how if the error was corrected Tycho would have come out with close to the same mean distance as the others; this explanation is omitted.

the same force. For the length of a pendulum oscillating in seconds, at the latitude of Paris, is three Paris feet $8\frac{1}{2}$ lines, as Huygens has observed. And the height which a heavy body traverses in falling in the time of one second, is to half the length of this pendulum, in the duplicate ratio of the circumference of the circle to its diameter (as Huygens has also pointed out). It is therefore 15 Paris feet 1 inch $1\frac{7}{9}$ lines. And because the force which holds the moon back in its orbit, if it should descend to the surface of the earth, comes out equal to our force of gravity, therefore (by Rules 1 and 2) it is that very force which we are accustomed to call gravity. For if gravity were different from it, bodies, in seeking the earth with the two forces conjoined, would descend twice as fast, and in falling in the space of one second would describe $30\frac{1}{6}$ Paris feet, in complete opposition to experience.

Notes on III.4

- **Note 1**

If you didn't happen to read the note entitled "Why Study III.4" (it appeared before Newton's statetment of this proposition), it would be helpful to do so before working through the expanded proof. Let it be incorporated here as Note 1.

- **Note 2**

This astonishing, wonderful demonstration deserves careful and orderly proof. Each part should be separately and completely argued so that when they fit together the full force of the evidence and thesis is appreciated.

The first part determines the accelerative force on the moon in orbit, and calculates how far the moon would fall in one minute of time if it were deprived of all tangential motion.

The second part determines the accelerative force that would be urging the moon at the earth's surface, assuming an inverse square force law, and calculates how far the moon would fall in one second at the surface.

The third part determines the accelerative force on a *rock* at the earth's surface, which we can measure most accurately using a pendulum. Here we call upon another unnumbered phenomenon. Using a proposition of Christian Huygens (see Note 7), Newton is able to calculate the distance a rock would fall in one second at the surface.

Finally we note that these latter two distances are almost identical, and by Rule 2, to the same effect we assign the same cause.

- **Note 3**

A passive read-through of this proposition will rob you of much of the excitement. It is important to work through all the calculations on your own calculator. Pause to ponder the coincidence—or try to explain it!—that gives the same number for the moon's fall in a minute from orbit and in a second at the surface.

And ask yourself whether you really would have believed (at least if you had not been taught so all your life) that the moon would fall as fast as a rock. Or is the amazing thing that it falls as slowly as a rock? What would even give someone the idea of dropping the moon and a rock together, or to suspect that the moon and a rock might obey the same laws of nature?

- **Note 4: Syzygies**

"Syzygies" means "places of being yoked together," namely, conjunction and opposition, when the sun, earth, and moon are in line.

In opposition, the moon is on the opposite side of the earth from the sun, and is seen as full. In conjunction, the moon is on the same side of the earth as the sun, and is seen as "new."

We need some particular place to take our measurements. Syzygy is a place popular with astronomers because that's where eclipses occur. (Among the useful opportunities offered by eclipses is the chance to observe the moon at conjunction.) This means that distances at syzygy had been measured much more often and much more carefully than at other places on the orbit. Lunar theories reflected this: some (most notably Ptolemy's) gave wildly erroneous distances at the quadratures, though they were nearly correct at the syzygies. So if Newton had not specified that distances were to be taken at syzygies, he would not have found anything approaching consensus on the mean distance.

- **Note 5: Some Useful Numbers**

Mean distance of the moon from the earth = radius of the moon's orbit = 60 earth radii.

$R = 60r$

Period of the moon's revolution against the fixed stars:

$P = 27^d\ 7^h\ 43' = 39{,}343'$

Circumference of the earth:

$c = 123{,}249{,}600$ Paris feet

Diameter of the earth:

$c = 2\pi r = \pi d$
$d = c/\pi$
$d = 123{,}249{,}600/\pi$
$d = 39{,}231{,}500$ Paris feet

Circumference of the moon's orbit:

$C = 2\pi R$
$R = 60r$
$C = 2\pi(60r) = 60(2\pi r)$
$c = 2\pi r$
$C = 60c = 60$ earth circumferences

$C = 60 \times 123{,}249{,}600$
$C = 7{,}394{,}976{,}000$ Paris feet

Diameter of the moon's orbit:
$D = 60d = 60 \times 39{,}231{,}500$ Paris feet
$D = 2{,}353{,}890{,}000$ Paris feet

Conversions

1 Paris foot = 0.3248 meters = 1.066 English feet
A "line" is a twelfth of an inch.
12 Paris lines = 1 Paris inch

- ● **Note 6: Timing the Fall of a Rock**

Why did Newton invoke inconvenient pendulum measurements and a pendulum equation to get the rate of fall of a rock at the earth's surface? Why didn't he just go out on the balcony and drop a rock and time it?

Newton and contemporary investigators lacked modern laboratory aids such as photo-gate timers or even stopwatches. In fact they had no clocks sufficiently accurate to time the fall of a rock with the precision Newton needs here for determining the acceleration of gravity.

But by using a pendulum it was possible to get the necessary accuracy. The investigator would time the movement of the sun over many degrees in many minutes of time while the pendulum was going through many oscillations. Then any error in the measurement of sun time would be spread out, with a very small error in terms of each second.

By adjusting the string length of the pendulum, the investigator could get it oscillating with a period of exactly two seconds. Then, using theorems derived for pendulums, conclusions about the force on the pendulum bob could be drawn with sufficient precision.

Newton cites the work of Christian Huygens, who performed many experiments with pendulums and formulated propositions on gravitational fall in his book *The Pendulum Clock*, published in 1673.

Huygens was the first to find the true value for the constant of gravitational acceleration, a number almost exactly the modern accepted value for his latitude. To do this he had to invent a clock accurate enough to do the measurements and he had to formulate an original set of mathematical theorems on both circular and gravitational motion for the calculations.

Expansion of Newton's Sketch of III.4

"That the moon gravitates towards the earth, and is always drawn back from rectilinear motion, and held back in its orbit, by the force of gravity."

To Prove:

The force of terrestrial heaviness and the centripetal force have the same effect at the same distance and may therefore be concluded to be identical.

Proof:

"The moon's mean distance from the earth in the syzygies, in terrestrial semidiameters, is 59 according to Ptolemy and most astronomers; 60 according to Wendelin and Huygens, $60\frac{1}{3}$ according to Copernicus, $60\frac{2}{5}$ according to Streete, and $56\frac{1}{2}$ according to Tycho."

See Note 4 for explanation of syzygies, and why the moon's mean distance was often calculated based on the distances measured there. Note that most astronomers, starting with Ptolemy, place it just around 60; Tycho Brahe, who gets a number significantly lower, is the exception.

"Let us assume that the mean distance is sixty semidiameters at the syzygies,..."

The final result of these considerations is the conclusion that the mean distance of the moon is 60 earth radii.

"...and that the lunar period with respect to the fixed stars amounts to 27 days, 7 hours, and 43 minutes, as is stated by the astronomers; and that the circumference of the earth is 123,249,600 Paris feet, as is established by the measuring Frenchmen."

Newton here gives us more numbers we will need to do the calculations of this proposition. There are a few we will need in addition, which can be derived from the ones Newton gives. Newton gives us the mean distance of the moon from the earth in earth radii, the period of the moon's revolution, and the circumference of the earth. We will also need the diameter of the earth, the circumference of the moon's orbit, and the diameter of the earth's orbit.

Note 5 gathers these in one list.

Part 1: First, We Calculate the Accelerative Centripetal Force on the Moon in Orbit.

"If the moon be supposed to be deprived of all motion and dropped, so as to descend towards the earth, under the influence of all that force by which (by Proposition 3 Corollary) it is held back in its orbit, it will in falling traverse $15\frac{1}{12}$ Paris feet in the space of one minute."

This is the statement of Part 1 of our three-step demonstration, the part that gives the accelerative force of the moon in orbit.

We will use a measure of this force, namely how far the moon will fall starting from the distance of its orbit in a given time, here one minute.

"This conclusion comes from a computation based either upon Proposition 36 of the first Book or (what amounts to the same thing) the ninth Corollary of the fourth Proposition of the same Book."

I.36 does get to exactly the conclusion of I.4 Corollary 9 by a more complicated route, deriving it as an exploration of consequences from things he develops later about conics. We will use I.4 Corollary 9, which we have proved.

"For the versed sine of that arc which the moon in its mean motion describes in the time of one minute at a distance of sixty terrestrial semidiameters, is about $15\frac{1}{12}$ Paris feet, or more accurately, 15 feet 1 inch and $1\frac{4}{9}$ lines."

Step 1: How far will the moon travel in orbit in one minute?

$$\frac{\text{distance of full orbit}}{\text{time of full orbit}} = \frac{\text{distance in one minute}}{\text{one minute}}.$$

$$\frac{7{,}394{,}976{,}000 \text{ Paris ft}}{39{,}343 \text{ minutes}} = \frac{x}{1 \text{ minute}}.$$

$x = 187{,}961$ Paris feet.

Distance traveled in orbit in one minute = 187,961 Paris feet.

Step 2: If deprived of tangential motion, how far would it fall in the same time under the influence of the same centripetal force?

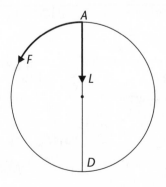

By I.4 Corollary 9, if the moon were released from a point in orbit, and fell towards the earth under the same force that held it in orbit, it would travel a distance such that:

$$\frac{AL}{\widehat{AF}} = \frac{\widehat{AF}}{AD}$$

that is,

$$\frac{\text{distance would fall } AL}{\text{distance traveled in orbit } \widehat{AF}} = \frac{\text{distance in orbit } \widehat{AF}}{\text{diameter of orbit } AD}.$$

For one minute:

$$\frac{AL}{187,961 \text{ Paris feet}} = \frac{187,961 \text{ Paris feet}}{2,353,890,000 \text{ Paris feet}}.$$

$AL = 15.0089161$ Paris feet, or 15 Paris feet, 0 inches, $1\frac{3}{11}$ lines.

This is the distance the moon would fall in one minute under the force that holds the moon in orbit.

This measures the accelerative force on the moon in orbit.

Conclusion of Part 1.

But wait! We have a problem here. The accelerative force on the moon in orbit may be more than just the accelerative force exerted towards the earth: what if the moon is also being accelerated towards the sun? Our measurement of the time it takes to move around its orbit would reflect both accelerations, and we really want just the acceleration towards the earth.

In fact, Newton is one step ahead of us in this, and the reason his number, 15 Paris feet 1 inch $1\frac{4}{9}$ lines, differs from ours, 15 Paris feet 0 inches $1\frac{3}{11}$ lines, is that he has taken into account the effect of the sun's force on the moon. He's been all through the detective story we're following, and knows how it comes out; and he can calculate, using Book I Proposition 45 Corollary 2, how much the sun perturbs the motion of the moon. It turns out that the action of the sun in drawing the moon away from the earth is to the centripetal force of the moon as 1 to $178\frac{29}{40}$.

He can also work his calculations backwards and subtract out the influence of the sun. So he multiplies the result (15.0089161 Paris feet per minute) by $178\frac{29}{40} / 177\frac{29}{40}$ to get 15.093366, or 15 Paris feet, 1 inch, 1.44 or $1\frac{4}{9}$ lines.

Without the sun drawing the moon away from the earth (at conjunction) and the earth away from the moon (at opposition), the moon would fall even farther in a minute than we observe it to do, by a little more than eight-hundredths of a foot, or more than an inch.

Part 2: Now We Consider What the Accelerative Force Would Be on the Moon at the Earth's Surface.

"Whence, since in approaching the earth that force increases in the inverse of the duplicate ratio of the distance, and is thus greater at the surface of the earth by 60×60 parts than at the moon, a body, in falling by that force in our regions, ought to describe a space of $60 \times 60 \times 15\frac{1}{12}$ Paris feet in the space of one minute, and in the space of one second, $15\frac{1}{12}$ feet, or more accurately, 15 feet 1 inch and $1\frac{4}{9}$ lines."

We're going to do a thought experiment here.

Let's assume that the centripetal force holding the moon in orbit obeys an inverse square law, as demonstrated in Proposition III.3.

If the relation $f \propto 1/r^2$ can be imagined to apply to different moon orbits at different radii, we could find different centripetal accelerative forces operating at those different distances.

So if we supposed the moon had an orbit with a radius twice as great as it currently has, we would expect the accelerative quantity of centripetal force acting on it there to be one quarter of what it is in its orbit of 60 earth radii.

But Newton challenges our imagination to bring it the other way. Suppose we brought the moon down to an orbit of just one earth radius. That is, imagine it to be in orbit just above the earth's surface. Then we will follow the same procedure we used in Part 1 and imagine it deprived of its tangential inertial motion, and let it just drop under the influence of that accelerative centripetal force. Just as we figured out how far it would drop in a given time from its orbit of 60 earth radii in Part 1, we can figure out how far it would drop in a given time from its new orbital distance of one earth radius.

$$\frac{\text{force at orbit}}{\text{force at earth's surface}} = \frac{(\text{radius at surface})^2}{(\text{radius at orbit})^2}.$$

$$\frac{f_o}{f_s} = \frac{r^2}{(60r)^2} = \frac{1}{60^2}.$$

Therefore the force should be 60^2 times greater at the earth's surface.

At this point in the proof, Newton tacitly relies on a relation he had proved in Lemma 10. Without going through the complete proof, we can make this relation plausible, as follows.

In *Two New Sciences,* Prop. 2 on accelerated motion, Galileo proved that

> "If a movable descends from rest in uniformly accelerated motion, the spaces run through in any times whatever are to each other as the duplicate ratio of their times; that is, are as the squares of the times."

That is,

$d \propto t^2$. (constant force)

This holds true if the accelerative force is the same. If, now, we consider two bodies being accelerated by different forces in the same times, Newton's

Definition 7 says that the accelerative forces are as the velocities generated:

$f \propto \Delta v.$

But for the same times, the distances travelled are proportional to the velocities. That is, if something is going two or three times as fast as something else for the same time, it will go twice or three times as far. Therefore

$f \propto d.$ (constant time)

If we allow both the forces and the times to be different, the distances will be proportional to the two conjointly—that is, we must compound (i.e., multiply) the ratios. So

$d \propto t^2 \times f$

Solving for f,

$f \propto d/t^2.$

   This proportionality depends on the forces being nearly constant, so it will only apply over short times and distances—times and distances for which the force will not vary. Now, our derivation of the distance of fall in Part 1 assumed a constant force in applying I.4 Corollary 9. So from that perspective, we need not be concerned about the force varying. But because we're doing a thought experiment, it is worthwhile to see whether the assumption of a constant force would be justified here.

   For this we appeal to experience: the acceleration of gravity at the earth's surface does not vary measurably over, say, 15 feet. At a distance of 60 earth radii the proportional distance would be 15 × 60 or 900 feet. Then if we are working with an inverse square law, a 900-foot change of altitude at 60 earth radii would vary the gravitational force only $1/60^2$ as much as does a 15-foot change of height at the earth's surface. Thus the conditions for use of the proportionality hold even better at the height of the moon than they do at the surface of the earth.

$F \overset{ult}{\propto} d/t^2.$

But we also know that $f \propto 1/r^2.$ [III.3]

Therefore

$d/t^2 \overset{ult}{\propto} 1/r^2.$

$d \overset{ult}{\propto} t^2/r^2.$ (1)

Let d_m be distance fallen at moon's orbit height

Let d_e be distance fallen at earth's surface height

Let t_m be time of fall at moon's orbit height

Let t_e be time of fall at earth's surface

Rewriting Equation 1 as a proportion,

$$\frac{d_m}{d_e} \stackrel{ult}{=} \frac{(t_m^2/r_m^2)}{(t_e^2/r_e^2)} . \tag{2}$$

Because Newton uses minutes at the moon's orbit height but converts to seconds at the earth's surface here, we must reduce the time by a factor of 60.

$$t_m = 60 \times t_e.$$

And we must reduce the radius by 60 to go from the moon's orbit radius to the earth's radius.

$$r_m = 60 \times r_e.$$

Substituting into Equation 2:

$$\frac{d_m}{d_e} \stackrel{ult}{=} \frac{(60t_e)^2/(60r_e)^2}{t_e^2/r_e^2}.$$

$$\frac{d_m}{d_e} \stackrel{ult}{=} \frac{(60)^2(t_e)^2/(60)^2(r_e)^2}{t_e^2/r_e^2} .$$

$$\frac{d_m}{d_e} \stackrel{ult}{=} \frac{t_e^2/r_e^2}{t_e^2/r_e^2} .$$

Since the second ratio of the proportion is a ratio of equality, so is the first, and

$$d_m = d_e.$$

In one *second*, the moon should fall 15 Paris feet, 1 inch, $1\frac{4}{9}$ lines at the surface of the earth, just as it falls 15 Paris feet, 1 inch, $1\frac{4}{9}$ lines in one minute at the distance of the moon's orbit.

This is what we would expect, at any rate, if the same accelerative centripetal force that is now holding the moon in its orbit would still be operating on it if it were brought to a point (a very low imaginary orbit) close to the surface of the earth.

Part 3: What is the Accelerative Force on a *Rock* at the Earth's Surface?

"And heavy bodies on earth do in fact descend with the same force."

We will show this using a pendulum bob at the surface of the earth. See Note 6 on the use of the pendulum for accurate timings for falling bodies.

"For the length of a pendulum oscillating in seconds, at the latitude of Paris, is three Paris feet $8\frac{1}{2}$ lines, as Huygens has observed."

We are given an observation, an unnumbered phenomenon, that at least at one point of the surface of the earth, near sea level, the string length associated with a pendulum whose period is two seconds is 3 Paris feet, $8\frac{1}{2}$ lines.

Converting to decimal, $L = 3.059$ Paris feet.

"And the height which a heavy body traverses in falling in the time of one second, is to half the length of this pendulum, in the duplicate ratio of the circumference of the circle to its diameter (as Huygens has also pointed out). It is therefore 15 Paris feet 1 inch $1\frac{7}{9}$ lines."

$$\frac{d}{L/2} = \frac{(2\pi r)^2}{(2r)^2} = \pi^2 \,.$$

$$d = \frac{L}{2}\pi^2 \,.$$

Since L was 3.059 Paris feet,

$d = 15.0956$ Paris feet.

Thus in one second a rock at the surface of the earth would fall 15.096 Paris feet, or 15 feet, 1 inch, $1\frac{7}{9}$ lines.

(In English feet, that comes out to 16.09 feet in one second, in case you want to compare it to the modern value.)

"And because the force which holds the moon back in its orbit, if the moon were to descend to the surface of the earth, comes out equal to our force of gravity, therefore (by Rules 1 and 2) it is that very force which we are accustomed to call gravity."

The moon at the surface of the earth, under the centripetal force that pulls it into orbit, would fall (by Part 2) 15 Paris feet, 1 inch, $1\frac{4}{9}$ lines in one second.

A rock at the surface of the earth, under the force of terrestrial heaviness, would fall (by Part 3) 15 Paris feet, 1 inch, $1\frac{7}{9}$ lines in one second.

By Rule 2, to the same effect, assign the same cause.

Therefore since by Rule 2 we would think that the moon would fall under the same accelerative force as a rock, the moon falls by gravity (terrestrial heaviness).

Notice that Newton didn't invoke Rule 2 at the start to say "something is attracting the moon towards the earth and something is attracting rocks to the earth, so to the same effect assign the same cause." He made sure the effect was not just generally the same, but the same as close as Huygens could make measurements. Newton's use of Rule 2 is not careless or crude, but very well-considered and supported by a line of argument that inclines one to think, "Yes, to deny the same cause here runs counter to the way we reason."

Then, of course, we might invoke Rule 4 and say: "And we will hold to this conclusion from induction despite any hypothesis that may be devised to circumvent it." We might need to do this because certainly someone could argue that lunar matter really is different, and if brought to the surface would fall $15\frac{1}{12}$ feet by centripetal force, but not at all by terrestrial heaviness, since it isn't terrestrial. And a rock brought to the distance of the moon might even fall $15\frac{1}{12}$ feet in a minute by terrestrial heaviness, but not be affected by the centripetal force that governs the celestial bodies.

"For if gravity were different from it, bodies, in seeking the earth with the two forces conjoined, would descend twice as fast, and in falling in the space of one second would traverse $30\frac{1}{6}$ Paris feet, in complete opposition to experience."

Once we have figured out, in Part 2, that the moon, if its orbit were brought down near the surface of the earth, would fall by centripetal force about $15\frac{1}{12}$ feet in a second, and have determined by Part 3 that a rock would fall the same distance in the same time by its terrestrial heaviness, if we weren't willing to say that the two forces are the same force, we would have to suppose that the moon would fall twice as fast, under the influence of both the centripetal force and its terrestrial heaviness.

In fact, we don't have experience of this, since we have not brought the moon to that low orbit and dropped it except in our thought experiment.

Newton may be suggesting here a thought experiment going the other way, in which we take bodies known to have terrestrial heaviness and imagine taking them up to the moon's orbit. Then presumably they would be moved by centripetal force just as the moon is. Then if we brought them back down we would have the same rocks our experience tells us fall only at a speed corresponding to one of the accelerations.

The accumulated proofs of this proposition have thus led us to conclude that the moon gravitates towards the earth by the force of terrestrial heaviness, as the enunciation stated.

Q.E.D.

Scholium

The demonstration of the proposition can be more amply displayed thus. If many moons were to revolve around the earth, exactly as in the system of Saturn or Jupiter, their periodic times (by an argument of induction) would observe the law of the planets discovered by Kepler, and therefore, their centripetal forces would be inversely as the squares of the distances from the center of the earth, by

Proposition 1 of this Book. And if the lowest of these were small, and were nearly to touch the peaks of the highest mountains, its centripetal force, which keeps it in its orbit, would (by the foregoing computation) be very nearly equal to the gravities of bodies on the peaks of those mountains. It would thus come to pass that if the same small moon were deprived of all the motion by which it proceeds in its orbit, it would descend to earth because of the loss of the centrifugal force by which it had remained in its orbit, and would do this with the same velocity with which heavy bodies fall on the peaks of those mountains, because of the equality of the forces with which they descend. And if that force by which that small lowest moon descends were different from gravity, and that small moon were also heavy towards the earth in the manner of the bodies on the peaks of the mountains, the same small moon under the two conjoined forces would descend twice as fast. Therefore, since both forces—the latter ones, of the heavy bodies, and the former ones, of the moons—look to the center of the earth, and are similar and equal to each other, these same [forces] (by Rules 1 and 2) will have the same cause. And consequently, that force by which the moon is kept back in its orbit, will be that very force which we usually call "gravity," and this must above all be true lest the small moon at the peak of the mountain either lack gravity, or fall twice as fast as heavy bodies usually fall.

Note on Scholium After III.4

It seems Newton still wasn't content that he had said enough that the force of Rule 2 would compel us to the conclusion that the centripetal force was gravity. So far from crudely relying on Rule 2, he wants the inductive step to be inescapable.

In addition to appreciating his efforts to satisfy us that we have a rigorous demonstration, we may be grateful for his scruples because they have given us this really delightful thought experiment of the little moons orbiting just at the mountain peaks: a charming finish to an exhilarating proposition.

General Scholium

The hypothesis of vortices is pressed by many difficulties. In order that any individual planet describe areas proportional to the time by a radius drawn to the sun, the periodic times of the parts of the vortex ought to have been in the duplicate ratio of the distances from the sun. In order that the periodic times of the planets be in the sesquiplicate ratio of the distances from the sun, the periodic times of the parts of the vortex ought to have been in the sesquiplicate ratio of the distances. In order that the smaller vortices about Saturn, Jupiter, and other planets be preserved in their circulations and float undisturbed in the sun's vortex, the periodic times of the parts of the solar vortex ought to have been equal. The rotations of the sun and the planets about their axes, which ought to have been consistent with the motions of the vortices, are in disagreement with all these ratios. The motions of the comets are in the highest degree regular, and observe the same laws as the motions of the planets, and cannot be explained by vortices. Comets are carried in highly eccentric motions into all parts of the heavens, which cannot happen unless vortices be removed.

Projectiles, in our air, feel only the resistance of our air. When the air is removed, as happens in Boyle's vacuum, the resistance stops, inasmuch as a slender feather and solid gold fall with equal velocity in that vacuum. And the account of the celestial spaces, which are above the sphere of the earth's exhalations, is the same. All bodies ought to move with complete freedom in these spaces, and for that reason the planets and comets revolve perpetually in orbits given in shape and position, following the laws set forth above. Though they will indeed carry on in their orbits by the laws of gravity, they nevertheless could by no means have attained the regular position of the orbits through these same laws.

The six principal planets revolve about the sun in circles concentric upon the sun, in the same direction of motion, approximately in the same plane. Ten moons revolve about the earth, Jupiter, and Saturn, in concentric circles, in the same direction of motion, very nearly in the planes of the orbits of the planets. And all these regular motions do not have their origin from mechanical causes, inasmuch as comets are carried freely in highly eccentric orbits, and to all parts of the heavens. In this kind of motion, the comets pass through the orbits of the planets with greatest ease and swiftness, and at their aphelia, where they are slower and delay for a longer time, they are at the greatest distance from each other, so that they pull each other least. This most elegant arrangement of the sun, planets, and comets could not have arisen but by the plan and rule of an intelligent and powerful being. And if the fixed stars be centers of similar systems, all these, constructed by a similar plan, will be under the rule of *One*, especially because the light of the fixed stars is of the same nature as the light of the sun, and all the systems send light into all mutually. And so that the systems of the fixed stars should not fall into each other mutually, he will have placed this same immense distance among them.

He governs everything, not as the soul of the world, but as lord of all things. And because of his dominion, he is usually called "Lord God Παντοκράτωρ." [1]

For "God" is a relative word, and is related back to servants, and "deity" is the absolute rule of God, not over his own body, as those believe for whom God is the world soul, but over servants. God most high is a being eternal, infinite, absolutely perfect; but a being without dominion, however perfect, is not the Lord God. For we say, "my God," "your God," "God of the Israelites," "God of gods," but we do not say, "my eternal," "your eternal," "eternal of the Israelites," "eternal of gods;" we do not say, "my infinite," or "my perfect." These names have no relation to servants. The word "God" everywhere signifies [2] the Lord; but not every lord is God. The absolute rule of a spiritual being constitutes God: true [absolute rule constitutes] the true [God]; the highest [absolute rule constitutes] the highest [God]; sham [absolute rule constitutes] a sham [God]. And from true absolute rule it follows that the true God is living, intelligent, and powerful; from the remaining perfections, that he is the highest, or in the highest degree perfect. He is eternal and infinite, omnipotent and omniscient; that is, he endures from eternity to eternity, and is present from infinity to infinity. He reigns over everything, and knows everything that happens or can happen. He is not eternity and infinity, but eternal and infinite; he is not duration and space, but he endures and is present. He endures always, and is present everywhere, and by existing always and everywhere, he has established duration and space. Since any single particle whatever of space is *always*, and any single indivisible moment whatever of duration is *everywhere*, surely the maker and lord of all things will not be *never, nowhere*. Every sentient soul is the same indivisible person at different times and in the different organs of perception and motion. Successive parts are given in duration, coexisting parts in space, neither [is given] in the human person or in his thinking principle: much less so in the thinking substance of God. Every person, *qua* sentient thing, is one and the same person throughout life in each and every organ of perception. God is one and the same God always and everywhere. He is omnipresent not in *power* alone, but also in *substance*.

---

[1] [Newton's marginal note:] That is, "Universal Emperor."

[2] [Newton's marginal note:] Our Pocock derives the word "deus" from the Arabic word "du" (and in the oblique case, "di"), which signifies the Lord. And in this sense, princes are called "dii", Psalm 84:6 and John 10:45. And Moses is called "deus" of his brother Aaron, and "deus" of king Pharaoh (Exodus 4:16 and 7:1). And in the same sense the souls of dead princes used to be called "dii" by the people, but falsely, on account of the want of dominion.

For power cannot subsist without substance. In him all things are contained[3] and moved, but without mutual effects [*passio*]. God is not affected by the motions of bodies, and these do not experience any resistance from God's omnipresence. It is universally acknowledged that the highest God exists necessarily, and by the same necessity he is *always* and *everywhere*. Hence also, the whole is entirely similar to himself, all eye, all ear, all brain, all force of perceiving, understanding, and acting, but in a manner by no means human, in a manner by no means corporeal, in a manner entirely unknown to us. Just as a blind man has no idea of colors, so we have no idea of the ways in which God most wise perceives and understands everything. He is entirely void of all body and corporeal form, and therefore cannot be seen, nor heard, nor touched; nor ought he to be worshipped under the image of any corporeal object. We have ideas of its[4] attributes, but we do not have the least knowledge of what the substance of any object is. We see only the shapes and colors of bodies, we hear only sounds, we touch only the external surfaces, we smell only odors, and taste flavors: we have no cognition of the inmost substances by any sense or act of reflection, and much less do we have an idea of the substance of God. We have cognition of him only through his properties and attributes, and through the wisest and best structures and final causes[5] of things, and marvel because of [his] perfections, and further, we revere and worship [him] because of his dominion. For we worship as servants, and God without dominion, providence, and final causes is nothing different from fate and nature. From blind metaphysical necessity, which is absolutely the same always and everywhere, no variation of things arises. The whole diversity of created things according to places and times could only have arisen from the ideas and will of a being existing necessarily. Moreover, God is said by way of allegory to see, to speak, to laugh, to love, to hate, to desire, to give, to receive, to rejoice, to become angry, to fight, to devise, to establish, to build. For every account of God is taken from human things through a certain likeness, not indeed perfect,

---

[3] [Newton's marginal note:] This was the opinion of the ancients, such as Pythagoras (in Cicero, *On the nature of the gods* Book 1, Thales, Anaxagoras, Virgil (*Georgics* 4:220, and *Aeneid* 6:721), Philo (*Allegories*, beginning of Book 1), Aratus (*Phenomena*, at the beginning). So also the sacred writers, such as Paul (*Acts* 17:27–28), John 14:2, Moses (*Deuteronomy* 4:39 and 10:14), David (Psalm 139:7, 8, 9), Solomon (*1 Kings* 8:27), Job 22:12, 13, 14, Jeremiah 23:23–24. Moreover, idolaters used to make out that the sun, the moon, and the stars, people's souls, and other parts of the world, are parts of the highest god, and are therefore to be worshipped, but falsely.

[4] The word Newton uses here, 'eius', could be masculine or feminine or neuter; hence, it is not possible to tell whether Newton means God's attributes or those of bodies. In the context of the preceding sentence, this word would refer to God, but in the context of what follows, it would refer to bodies. Although the latter translation has been chosen, as being more consistent with the argument, the former is not impossible, and was adopted by Motte. —Translator's note

[5] "Final cause" is an Aristotelian term that denotes the end for the sake of which something happens. See Aristotle, Physics, II.3, 194b32.—Translator's note

but of a certain sort. And this much concerning God, to discourse of whom, at least from the phenomena, is the business of natural philosophy.

Hitherto I have set forth the phenomena of the heavens and of our sea through the force of gravity, but I have not yet assigned the cause of gravity. This force does indeed arise from some cause, which penetrates all the way to the centers of the sun and the planets, with no diminution of power, and which acts, not according to the quantity of the *surfaces* of the particles upon which it acts (as mechanical causes are wont to do), but according to the quantity of *solid* matter, and which acts at immense distances, extended everywhere, always decreasing in the duplicate ratio of the distances. Gravity towards the sun is compounded of the gravities towards the individual particles of the sun, and in receding from the sun decreases precisely in the duplicate ratio of the distances all the way to the orbit of Saturn, as is manifest from the planets' aphelia being at rest, and all the way to the aphelia of the comets, provided that those aphelia are at rest. The reason for these properties of gravity, however, I have not yet been able to deduce from the phenomena, and I do not contrive hypotheses. For whatever is not deduced from the phenomena is to be called a *hypothesis*, and hypotheses, whether metaphysical or physical, whether of occult qualities or mechanical ones, have no place in *experimental* philosophy. In this philosophy, propositions are deduced from the phenomena, and are rendered general by induction. Thus the impenetrability, mobility, and impetus of bodies, and the laws of motions and of gravity, came to be known. And it is enough that gravity really exists, and acts according to the laws set forth by us, and is sufficient [to explain] all the motions of the heavenly bodies and of our sea.

It would now be appropriate to add some remarks about a certain extremely subtle spirit pervading gross bodies and lying hidden in them, by whose force and actions the particles of bodies attract each other mutually at least distances, and stick together when brought into contact, and electrical bodies act at greater distances, both repelling and attracting neighboring corpuscles, and light is emitted, reflected, refracted, inflected, and heats bodies, and all perception is aroused, and the members of animals are moved by the will, that is, by vibrations of this spirit propagated through the solid filaments of the nerves from the external organs of perception to the cerebrum and from the cerebrum to the muscles. But these cannot be set forth in a few words, nor is there at hand a sufficient body of experiments by which the laws of action of this spirit are required to be accurately determined and shown.

Further Reading

Berlinski, David, *Newton's Gift.* The Free Press (Simon & Schuster), 2000.

Brackenridge, J. Bruce, *The Key to Newton's Dynamics.* University of California Press, 1995.

Christianson, Gale, *Isaac Newton and the Scientific Revolution.* Oxford University Press, revised edition 1998.

Cohen, I. Bernard, *Introduction to Newton's 'Principia'.* Harvard University Press, 1971.

Cohen, I. Bernard, and George E. Smith, *The Cambridge Companion to Newton.* Cambridge University Press, 2002.

Cohen, I. Bernard, and Richard S. Westfall, editors, *Newton: Texts, Backgrounds, Commentaries.* W. W. Norton & Co., 1995.

Crowe, Michael J., *Theories of the World from Antiquity to the Copernican Revolution.* Dover Publications, 1990.

Crowe, Michael J., *Mechanics from Aristotle to Einstein.* Green Lion Press, publication date 2005.

De Gandt, François, *Force and Geometry in Newton's Principia.* Princeton University Press, 1995.

Densmore, Dana, *Newton's Principia: The Central Argument.* Green Lion Press, Third Edition, 2003.

Fauvel, John, et. al., *Let Newton Be!* Oxford University Press, 1988.

Galileo, *Two New Sciences,* translation by Stillman Drake. Wall & Thompson, 1989.

Gleick, James, *Isaac Newton.* Pantheon Books, 2003.

Guicciardini, Niccolò, *Reading the Principia.* Cambridge University Press, 1999.

Newton, Isaac, *Principia,* I. Bernard Cohen and Anne Whitman, translators. University of California Press, 1999.

Westfall, Richard S., *Never at Rest: A Biography of Isaac Newton.* Cambridge University Press, 1984.

Wilson, Curtis, *Astronomy from Kepler to Newton.* Variorum Reprints, London, 1989.

Yoder, Joella G., *Unrolling Time.* Cambridge University Press, 1988.

Selections from Newton's Principia:
a science classics module for humanities studies
is adapted from the book
Newton's Principia: The Central Argument.

from the reviews

"Densmore's goal is to help students comprehend Newton's demonstrations in their own terms. The aim is not to tell students what Newton demonstrated, but to enable them to understand the force of the demonstrations by repeating them. Although attention remains focused on mathematical demonstrations, the attitude is not ahistorical in any way. Rather, we stand as it were at the historical moment when Newton first elaborated the demonstrations and scientific thought attained a new level of understanding. ...

"This is a wonderful book. Taking Newton in his own terms, it insists on the full rigor of the demonstrations and does not hesitate to point out where full rigor appears to be lacking. ... As she says in the preliminaries, 'we understand Newton only in understanding why he proved things as he did.' Students are not the only ones who can profit from the exercise."

Richard S. Westfall, author of *Never at Rest*. Review in *Isis*.

"The stress is on encouraging students to reconstruct Newton's proofs in their original geometric form, rather than translating them into the more familiar symbolic calculus. This is particularly interesting because Newton's geometric style informs our geometric and physical intuition in a way which is complementary to the understanding achieved via analytical tools. ... It is a pleasure to follow Densmore's reconstruction of this momentous discovery in science, since the argument supporting it requires on the one hand very elementary mathematical tools, and on the other a profound understanding of the relationships between mathematical models and astronomical data.

"Densmore's book is interesting not only for teaching purposes. Historians of science have a great deal to learn from it. The *Principia* is always difficult to read since Newton is often quite brief and leaves the reader to reconstruct the steps of the complete argument. This guidebook provides such an analysis. Densmore's book is a first-class work: it is a detailed, useful and enjoyable commentary on those mathematical demonstrations in which the theory of universal gravitation was first established."

Niccolò Guicciardini, review in *Math Reviews*

"Densmore's commentary has a directness, an intelligence and infectious energy that takes readers through all the difficulties to a very satisfying accomplishment...I cannot emphasize too strongly what an achievement it is."

Curtis Wilson, Editor,
General History of Astronomy

Newton's *Principia*: The Central Argument, Green Lion Press, third edition 2003.
ISBN 1-888009-24-1 cloth binding, ISBN 1-888009-23-3 sewn softcover